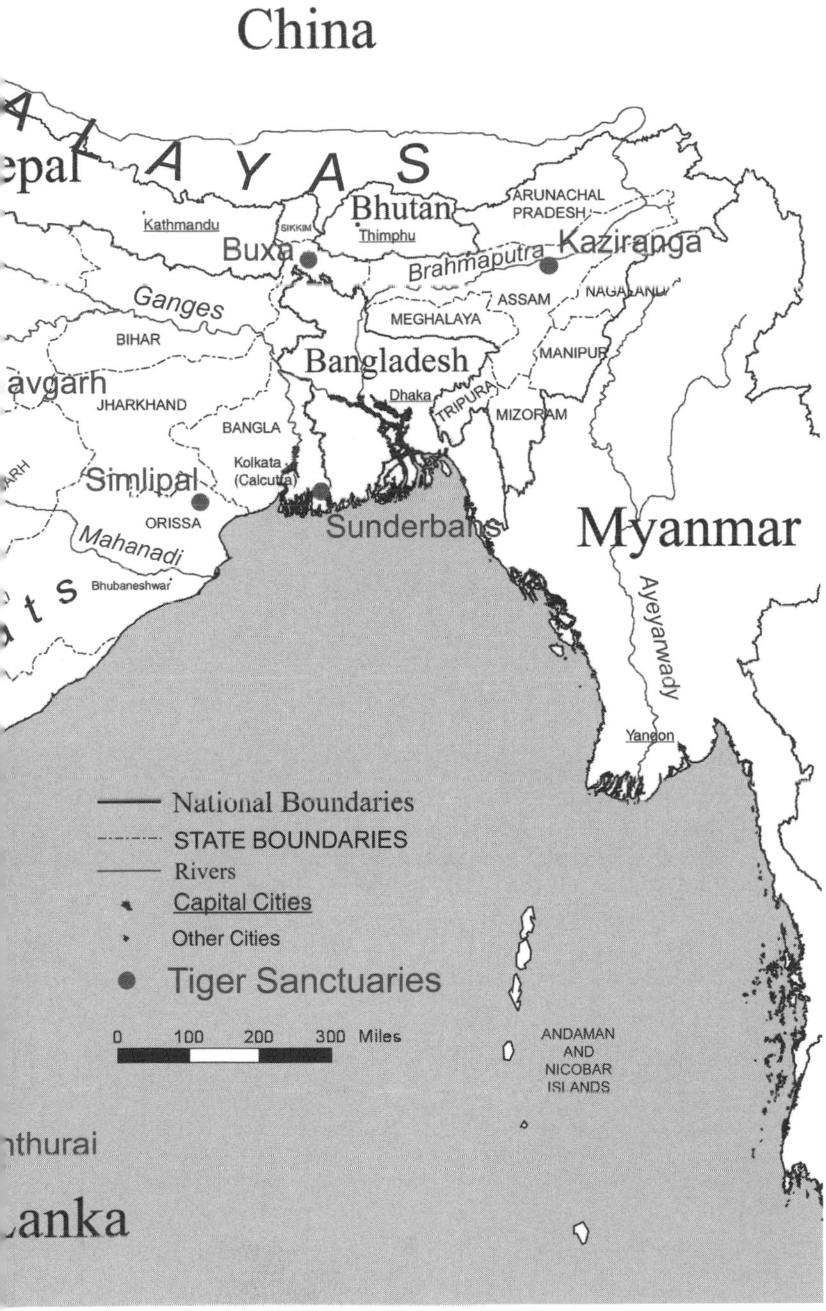

China

Nepal
ALAYAS
Bhutan
ARUNACHAL PRADESH
Kathmandu
SIKKIM
Thimphu
Buxa
Brahmaputra
Kaziranga
Ganges
ASSAM
NAGALAND
MEGHALAYA
BIHAR
Bangladesh
MANIPUR
avgarh
JHARKHAND
Dhaka
TRIPURA
MIZORAM
BANGLA
Kolkata (Calcutta)
Simlipal
ORISSA
Sunderbans
Myanmar
Mahanadi
Bhubaneshwar
ts
Ayeyarwady
Yangon

National Boundaries
STATE BOUNDARIES
Rivers
Capital Cities
Other Cities
Tiger Sanctuaries

0 100 200 300 Miles

ANDAMAN AND NICOBAR ISLANDS

nthurai
Lanka

THE LIFE AND FATE OF THE INDIAN TIGER

Tobias J. Lanz

Praeger
An Imprint of ABC-CLIO, LLC

A B C CLIO

Santa Barbara, California • Denver, Colorado • Oxford, England

Library of Congress Cataloging-in-Publication Data

Lanz, Tobias J.
The life and fate of the Indian tiger / Tobias J. Lanz.
p. cm.
Includes bibliographical references and index.
ISBN 978–0–313–36548–5 (hard copy : alk. paper) — ISBN 978–0–313–36549–2 (ebook)
1. Tigers. 2. Wildlife conservation. 3. Tigers—Habitat—India. I. Title.
QL737.C23L3635 2009
599.7560954—dc22 2009011684

13 12 11 10 09 1 2 3 4 5

This book is also available on the World Wide Web as an eBook.
Visit www.abc-clio.com for details.

ABC-CLIO, LLC
130 Cremona Drive, P.O. Box 1911
Santa Barbara, California 93116-1911

This book is printed on acid-free paper ∞

Manufactured in the United States of America

Contents

Introduction

The tiger is one of the world's most endangered animals, having been driven to near extinction throughout its native Asia. As recently as 2008, as few as 3,000 tigers were believed to exist worldwide, greatly diminished from a population of 100,000 eighty years ago. Amazingly, the majority of the world's tigers—some 1,500—still live in India, a country of one billion people crammed into an area one-third the size of the United States. India is notorious for poverty, disease, corruption, ethnic violence, terrorism, and rampant urban sprawl. The tiger should have vanished long ago. Yet it has managed to survive.

I came to India over five years ago to find out why. Originally, I believed the answer was because of good conservation projects and policies. After all, India is famous for Project Tiger, the pioneering conservation program initiated decades ago to save the cat. But, after traveling throughout the country, visiting India's many tiger sanctuaries and talking to people from every walk of life, I realized the tiger's survival was based on more than good conservation. The tiger has survived in India because of the country's unique synthesis of nature and culture.

India has the most diverse landscape of any country on earth. It ranges from the freezing Himalayas to scorching deserts and fertile plains, and it includes almost every type of forest and grassland imaginable. The adaptable tiger can thrive in all of these rich habitats, and often in large numbers. And so do people. Every place that has tigers also has people, creating intimate and unique relations between the two.

India's diverse geography also created the world's most fantastic array of religions, races, castes, and tribes. Each has its distinct traditions and its own experiences with tigers. As such, the cat is part of many different myths and legends. It is depicted in art and architecture, adorns jewelry and clothing, and is used as a religious and political symbol. It is a favorite advertising image and corporate logo. The cat is alternately feared and revered, loathed, praised, and imitated.

The goal of this book is to explore this complex topic—the life and fate of the Indian tiger (the world's best hope for tiger survival)—and share it with the average educated reader. But this requires more than just a description of social and natural history. It requires knowing how tigers behave and where they actually live. Tigers are forest animals. And they can survive, even thrive, if given enough cover and prey. They are highly adaptable and make their homes in some of the harshest conditions on earth, as well as the most verdant and beautiful.

I describe all these places in my travels to India's various tiger sanctuaries. I present the natural and social history relevant to each locale, the local geography, plant and animal life, and the many cultures encountered. I also delve into history to describe the lives of people who once shared their lives with tigers—from hunters and explorers to pioneering conservationists, religious seekers, and others.

Each chapter also provides social criticism and commentary on the current problems facing the tiger. I try to be fair in my assessments, though sometimes strident in my critique, as I cannot hide my sympathies for the cat, because the tiger's entire fate lies increasingly in human hands. So I scrutinize the role of government and conservation groups, business corporations, local villagers, and even organized crime and their impact on tigers—directly and indirectly, for better or worse.

The ultimate question this book tries to answer is whether the story of the Indian tiger will have a happy ending. Will the tiger survive in India? The cat is not safe everywhere in the country, and many wonder whether the long and intimate relationship between tigers and people is now coming to an end with continued population growth and growing prosperity. There are many places in India with no tigers at all, where they have been poached out of existence or overcome by the momentum of population. Other regions have a few, maybe dozens, remaining. Only a handful of sanctuaries have a hundred or more cats.

The situation is certainly troubling. Nevertheless, there are conservation victories, and many people fight to protect the cat. So the last chapter of the Indian tiger's story has not been written. India remains the last place on earth where the tiger has a fighting chance to survive. Conversely, it may be the last place a wild tiger is seen alive. Either way, India is where the fate of the species will be determined.

— 1 —

King of the Jungle

The lion is called king of the jungle. But it does not live in jungles. It is a savannah creature. The tiger is the real king of the jungle because it is the only big cat that lives exclusively in forests and needs forests to survive. It is a quintessential jungle creature. Naturally it has always thrived in India because India is the quintessential jungle land. It has more types of jungle than any other country on earth. Even the English word "jungle" is derived from the ancient Sanskrit term *jangli*, referring to the wildest forests, those filled with dangerous and mysterious beasts, especially tigers.

When tigers migrated into India thousands of years ago from eastern Asia, they found their perfect home.[1] Tigers thrived in the marsh and forest mosaic that grew along the great northern rivers, along with imposing herds of rhinoceros, water buffalo, and swamp deer. The big cat was equally at home in the dry interior, hunting deer and antelope in the wooded hill tracts. Tigers even adapted to India's eastern and southern rainforests, sharing the dense tangle of vegetation with elephants and primates. And in the impenetrable mangrove forests that straddle the eastern coastline, it was the tiger, not the leopard, that established itself as the sole and supreme predator. Tigers established themselves everywhere across the subcontinent except the far western deserts and the frigid Himalayan peaks.

The best tiger forests are a mix of woods and grasslands with ample water supplies. This habitat supports large game populations—especially deer, antelope, and wild boar—the tiger's main prey. These forests can support 20, even 40, tigers per 100 square miles. Today these ideal tiger forests are found only in India, which is why half the world's tigers live there despite the high human population. In contrast, the rainforests of Myanmar and Southeast Asia, also tiger havens, support only four or six cats per 100 square miles, and the vast boreal forests of the Russian far east, home of the Amur or Siberian tigers, support even fewer. Today India's tiger forests exist in hundreds of fragments, like so many jewels scattered across the land. Most are managed as government wildlife sanctuaries, national parks, or special tiger reserves. One of the largest and most diverse stretches of forest is in the Western Ghats (mountains) in southern India. It is where I first encountered the King of the Jungle.

Kalakad-Mundanthurai is India's southernmost tiger reserve, located at the very tip of the Western Ghats. It about 300 square miles in size, and, like most Indian wildlife sanctuaries, it is hemmed in by farms and villages. Kalakad has a dozen distinct forest types, ranging from dense rainforests to dry deciduous woods and scrub forests.

To reach Kalakad-Mundanthurai I spent a full day traveling across the flat plains of southern India. I first took a bus from the city of Madurai to Tirunalvelli, where I hired a taxi for the final leg. The Ghats soon came into view, but we skirted them for over an hour before finally turning towards a deep ravine where a narrow road wound its way into the hills.

The road rose steeply from the flat and populated plateau. It wound its way through scrub forests well suited to the rocky soil, the greenery occasionally broken by cactus patches or granite outcroppings left over from volcanic activity that ceased eons ago. The outcroppings became more frequent and imposing as our dilapidated taxi climbed higher. Some formed the backbone of small mountain ranges, while others jutted hundreds of feet above the surrounding forest. Either way the dark gray rocks were stained with white streaks of guano left behind by vultures and other raptors.

A small hand-painted picture of a tiger marked the sanctuary boundary. The faded sign stood a few yards in front of the inconspicuous entrance, the gate open and unattended. The air was cooler now, thanks to the higher elevation and the fact that the afternoon sun was blocked by the surrounding hills. The forest had also begun to change. Scrub trees remained dominant, but larger deciduous specimens now became common, especially around streams and water holes, their splaying leafy branches creating shady resting places.

I spotted gray langur monkeys (*Semnopithecus hypoleucos*) in a stand of large trees surrounding a pool of water. Their shy demeanor and graceful ways have endeared them to the Indian people for millennia, making this monkey an important Hindu deity throughout the land. The monkeys were accompanied by small herds of spotted deer (*Axis axis*) or chital, which ate the leaves and fruit that dropped to the ground during the monkey's frenetic feeding sprees. However, they have a symbiotic relationship that goes beyond food. The monkeys watch from the treetops as the deer browse and alert them with sharp cries should a tiger or other predator appear. And the deer reciprocate when the monkeys leave the safety of the treetops to rest or feed on the ground.

Bird life was equally rich. Mynahs and forest babblers flitted noisily about the underbrush, and tiny green bee-eaters darted out of the forest cover to snatch up insects along the road's edge. I could make out the silhouettes of birds as they glided between the distant treetops and could hear their muted calls echoing from the hillsides in chorus-like intervals. Unfortunately, the most famous Indian bird, the peacock, did not make an appearance that day. Its characteristic "may-haw" cry, together with the tiger's roar and the spotted deer's bark, are the quintessential sounds of wild India.

My driver rounded a few more turns until we came to a rest house built by the British in the late nineteenth century as the forest administrator's residence. Today this grand old building serves as the only tourist lodge in the entire reserve. The two-story structure with a bleached white exterior features an imposing balcony with nice views of the Thamirabarani River flowing lazily through the forest clearing below. With its vantage over the surrounding landscape, I hoped to perhaps catch a glimpse of a tiger. Instead I encountered bonnet macaques (*Macaca radiata*), a more aggressive and less attractive relative of the langur, who used the roof as a launching pad to climb into treetops or simply as a sounding board to annoy people passing below.

After settling into the rest house that evening, I strolled down one of the narrow forestry roads to survey the tiger's domain up close. As the sun set, it turned the autumn sky a deep orange, illuminated the thorny branches of the scrub forest, and cast long shadows across the road. As the shadows grew to darkness, the forest came alive with night sounds. I knew that my chances of encountering a tiger were slim because they prefer the more secluded interior; nonetheless, walking through these tiger forests unarmed in the dark was both unnerving and exhilarating.

As I walked along the road, alert to every sound and movement from the surrounding brush, I tried to imagine what such a walk would have been like a hundred, or even a thousand, years ago. A tiger bellowing in a nearby forest or springing suddenly and silently out of a thicket to take down a deer or wild boar was commonplace. For centuries contact between tigers and people was frequent because tigers were everywhere. The relationship was usually one of respect, but confronting a 400-pound cat that could kill in seconds had to be heart stopping. In the past, tigers killed people and people also killed tigers. This subtle balance of power was the law of the forest for millennia.

I turned back toward the rest house as the evening grew darker. A few birds darted into the safety of the forest to sleep. I stopped once, when I heard the undergrowth rustle. I stood for a second and then moved on when I realized it was some small creature, perhaps a squirrel or even a deer. Stars were now appearing and a half moon hung low on the horizon. I rounded a bend and the rest house came into view, its lone porch lamp creating the only light in an otherwise pitch-black forest.

That night, I lay awake in the old lodge deciphering the forest sounds—the constant whir of insects and the occasional call of a night bird. I wondered what it was like when a colonial officer lived here. His only regular human contact would have been with hunting and gathering tribes of the forest interior and the farming communities in the outlying plains. Yet his solitary life had privileges, for he experienced the true Indian jungle, in which the tiger's roar was surely heard. I too, listened intently, but the tiger did not roar that night.

◆ ◆ ◆

The next day I went to an ancient temple called Thembarvandi, located a few miles away from the rest house but still inside the sanctuary. Like most Hindu temples it was built in a place of natural beauty and mystery, often a mountaintop or deep in the forest, in this case next to a series of rapids on the Thamirabarani River that flowed over large volcanic plates spread out for several hundred feet.

The temple sat upon a ledge overlooking the water. It was hewn of the surrounding granite, of which it still seemed to be a part. Although simple in design, it was evocative, gathering into itself and reflecting the surrounding clear stream, blue sky, and green hills. Almost every tiger sanctuary in India has a temple or shrine, which is only fitting as wild places always evoke the sublime. Most, like Thembarvandi, are ancient and still have loyal attendance. A few dozen people were praying and bathing here that day, much like their ancestors had done for centuries.

My driver explained that this place draws over 200,000 people during religious festivals. They are pious and reverent and do not mean to harm forests or tigers. But with such a crowd, destruction is inevitable. There are no facilities, so trees are cut for cooking, streams become toilets, and branches shelter. The filth and destruction left behind is overwhelming.

Despite the obvious problem, it is difficult to overcome. The government does not have the resources to manage such massive events, nor can it ever hope to curtail a practice so deeply rooted in history and local life. It clearly illustrates how the momentum of tradition overwhelms a modern idea like conservation.

I noticed the areas close to the river had been stripped of vegetation and were surrounded by trampled soil that looked like a livestock pen, the result of the festival months earlier. It is one of the paradoxes of tiger conservation. These people, who have lived close to India's forests and its tigers for so long, are now becoming one of the threats to their survival.

After we left the temple, my driver suggested we spend the night in the nearby village, also home to the local zamindar, a minor aristocrat, whose family had been guardians of the temple since its beginning and even today help initiate the yearly rituals that draw the faithful. The zamindar was also known as a keen conservationist, as much of his family's one-time hunting grounds are now part of the tiger sanctuary.

The zamindar lived in a village with a couple hundred residents and few Western amenities. The huts were typical mud and brick with thatched roofs. Cows and goats trampled through muddy streets, some of which had been strewn with straw to give the occasional car or motorcycle some traction.

The zamindar's palace was a modest structure erected in the center of town. Although less than a hundred years old, it was perfectly medieval in design. Its fortress-like walls blended with the drab brown soils of the village, and there were two huge wooden entry doors, each with massive steel doorknockers.

I wanted to meet the man, so we parked and I walked up to the double doors. After several minutes of heavy pounding, the doors finally opened and a servant greeted us. After we explained our intent, he graciously asked us inside and led us towards the zamindar's sitting room.

The dark and spacious entryway featured walls adorned by huge black and white photographs, faded with the passage of time and warped by the moist tropical air. Family portraits were intermingled with hunting shots, showing the invincible hunter posing triumphantly with his lowly and unlucky prey. Although disconcerting to the modern eye, these images provided a powerful insight into the relations between man and beast from this bygone age. Through narrow windows, I could see the palace outside with its labyrinth of hallways, courtyards, and gardens connecting its many rooms, like a village within a village, one that comfortably housed the royal family, their servants, guests, and a number of livestock.

As I glanced back at the pictures, the zamindar entered from the far hallway. He was an elegant man in his late sixties, thin with a slight mustache and slicked-back gray hair. He wore a Western-style collared shirt and a pair of traditional *lungies*. We exchanged greetings, and I asked if we could talk about tigers. His eyes lit up and then he frowned. "There are only a few left here, you know."

I pointed at the photographs and asked why there were no tigers among all these trophy elephants, bears, and bison. He said his family never hunted the cat. Only problem tigers like "cattle lifters" and man-eaters were shot. Other zamindars certainly did kill tigers, but his family never found much sport in it. "Besides, they are hard to hunt in this hilly terrain. You never see them," he laughed.

His family once owned tens of thousands of acres of forest and farmland and ruled over several dozen villages. Their land had been reduced to 2,000 acres today and a half dozen villages. The remaining lands were taken by the state or privatized after Indian independence in 1947. The forests became government reserves, which allowed timber harvesting and grazing. Since the expropriation of these lands in the early 1950s, the zamindar has spent much time secluded in his palace, and for the past few decades, reading and writing his views on a number of topics, most of which center on the passing of the old order.

Like many Indian aristocrats, he was educated and trained as an "English gentleman." He learned European history, literature, table manners, and, of course, marksmanship. As we walked up two flights of stone steps to his sitting room, he told me how he loved to be out in the forests as a young man. He went there to hunt or simply to enjoy the solitude. He pointed out the window to the distant farms. "As a child I remember the forest covered half the valley and we often saw wild animals come out to the edge of our fields. Tigers used to come out of the woods in broad daylight completely unafraid of humans. Only man-eaters harmed people. But they were rare. Tigers helped us because they kept those destructive pigs and deer under control."

We sat down in a small study that adjoined his sitting room. Here the zamindar spent much of his day writing and reading. I asked him about his own family and his life. He explained that his family had lived here for a thousand years.

"The human population was much smaller when I was young. There was so much forest then and we all lived from the forest. Today all the trees are gone except those high in the hills. That is where the tigers hide. The forests have all been cut down for farms and pastures." He pointed back out of the window again to some low barren hills, where a few cattle grazed among rock outcroppings. "See there. That was all once forest. Look at it now. This is how the forest looks everywhere." Then he gave a sigh of resignation. "But you cannot just blame these simple farmers," he added, "The government is also to blame for what happened to the tiger and the forest. Government officials are corrupt. They plunder the forest for their own needs. When officials are posted here to take care of the forest they use it for their own gain. They do not care about the forest or the people. You see, the people know this and that is why there is no respect." He thought the idea of government wildlife sanctuaries was good, but the government "botched" it.

He leaned back in his chair and waved his finger in the air. He came close to me, looked me in the eye, and spoke quietly, as if confiding in me. "The government now has the power here. But, you know, if I wanted to raise an army of 50,000 men in this valley I could still do it!" When I asked if they would die for him, he exclaimed, "Of course, I am their natural leader." And when I asked whether a government bureaucrat could evoke such loyalty and obedience, he scoffed, "Never," with a raised voice. "The people do not trust them. They are scoundrels."

We were ready to go downstairs for dinner when drums, horns, and chanting sounded in the entry hall below. Four young men proceeded up the stairs into the sitting room, stood before the zamindar, and then stopped the music. He gestured as if to give a blessing and then spoke a few words in his native Tamil. The men bowed and left silently. The zamindar explained these were villagers who came to honor him. They did this every evening. And it had been done for a thousand years.

I left the zamindar and his village the next day before noon. We drove down the dirt lane past the tiny earthen huts and sagging stone buildings, a tremendous contrast to the large palace behind us. And it was this social and economic disparity that the Indian government had tried to rectify when it expropriated the zamindar's and other royal lands after independence. It was part of the march towards democracy, equality, and freedom. Indeed, many people were freed from the hard life under the ancien régime.

After talking to the zamindar, I began to understand another side to the old order—the subtle and fragile bond between lord, peasant, forest, and tiger. By destroying aristocratic rule, the Indian government and other modernizing forces had unraveled the spiritual connections that sustained traditional society and bound it to nature. As a result, any balance or harmony that may have existed between people, animals, and their common forest homes also came undone.

There was another connection. The aristocrat stood at the top of human kingdom, just as the tiger stood atop the animal kingdom. But the two were always intertwined, always connected to the forest for well-being and survival. It was no accident that the demise of the aristocrat and the tiger coincided. And both had to retreat into isolation—behind the walls of the fortress or to the depths of the forest—to save themselves from the seismic political, economic, and technological shifts of the modern era.

Ironically, the bureaucrat and the scientist, vanguards of modern revolution, now struggle to preserve artifacts of antiquity and wilderness areas—protecting them against the very forces they unleashed.

◆ ◆ ◆

From Kalakad I traveled to Periyar Tiger Reserve in the neighboring state of Kerala, a good day's drive. Once called the Malabar Coast, the region was renowned for its luxuriant tropical forests, home to elephants, primates, and tigers. They were rich in timber, especially teak, and spices. And many of the world's most common spices, like black pepper, turmeric, cardamom, and cinnamon, are native to the region. Today it remains a major producer of spices and tea, lucrative crops that make Kerala one of India's wealthier states.

This forest wealth attracted traders and planters from Europe, the Middle East, and Southeast Asia for centuries—in fact creating a cosmopolitan culture well before the British rule. Today a large percentage of the population is Muslim and Christian. There is even a small Jewish colony, now fading, that was established 2,000 years ago by believers seeking religious freedom from the Roman Empire.

Periyar is tucked into the heart of the Western Ghats rainforests, and it is one of a few Indian tiger sanctuaries with rainforest predominating. But it is unusual because it has a higher tiger population than most rainforests, around 40, because of the grasslands that are abundant in the higher elevations and around a reservoir built here a hundred years ago—all of which support good prey populations, hence more tigers.

The lake and the open grasslands make wildlife viewing easy, which has made Periyar the major tourist destination in southern India. People can see elephants, deer, forest bison (or gaur), monkeys, and occasionally the elusive tiger from tour boats and guided hikes on forest trails. I had heard about a recent daytime tiger sighting not too far from the forest lodge where I was staying, so I hired a guide one morning and trekked out to the location to see if lightning would strike twice.

We wandered through dense rainforest for a good while. The treetops, a contiguous mass of intertwined branches and leaves, teemed with bird and insect life. But the interior was dark and gloomy; gray mosses and brilliant green ferns covered tree trunks whose open cavities were colonized by masses of orchids. Huge vines the thickness of a man's arm hung low across the empty chasms formed by trees that towered a hundred feet overhead.

The forest floor was littered with leaves and rotting logs. Termite mounds, black as the soil and hard as cement, were common. Some looked like mushrooms with a "hat" atop a thick stem, whereas others were plastered to the sides of tree trunks. Occasionally tiny colorful butterflies flitted by and disappeared into the shadows. The most common creature was the leech. Hundreds made their way across the ground towards any moving object with great speed and precision. We stopped every 20 minutes to pull them off shoes and pants, making sure none attached themselves with their rasping mouths. Their bite is painless, allowing them to attach themselves and swell to the size of a grape before falling off or being discovered by an incredulous and sometimes hysterical host.

The forest path itself was narrow and well worn, used by animals and people for decades, maybe longer. Paths like these once crisscrossed all of India's forests. It is hard to know who made them first, animals or people. These simple dirt lanes, which cut through the densest woods, were important to both tigers and people as both depended on them for their survival. They connected distant villages to one another, allowing trade, communication, and a way for religious pilgrims to attend temples. They helped tigers to survey their territories, access game, and look for mates. Tigers follow specific trails, marked by scent, which establishes their hunting and mating territories.

Powerful males have the largest and best territories—sometimes covering hundreds of square miles, depending on forest type and prey densities. Female territories are smaller and several will overlap with a dominant male's. Young males have the smallest territories, usually on the forest periphery, where they often come into contact with villages and farms.

The trail took us to the clearing where the cat had been spotted. It was a good tiger habitat—a grassy area where deer and boar congregated, surrounded by dense thickets where the cat could lie in ambush. I looked across the meadow to the distant woods and saw a slight rustle. I waited. Maybe a tiger. Out bolted a shaggy black shape. It was a large male boar, probably 200 pounds, with a heavy mane and gleaming white tusks.

Boars are favorite tiger food, but a male this size could prove too much even for the big cat, as it can inflict serious damage or even death with his sharp tusks. The boar looked about furtively and then scampered back into the woods. Despite his armature and size, the boar is also prudent as the agile tiger still wins most encounters.

The tiger's colors are key to its hunting success. When seen in a zoo or picture, the bold orange and black stand out. In the forest it is the opposite. Black and yellow stripes mimic sunlight and shadow, shifting the eye's focus away from the tiger's body to its surroundings. In tall grasslands, the cat blends perfectly with the browns, yellows, and dull greens. It disappears just as easily into the interplay of light and dark in a shaded forest. Even when crouched in an open ravine, the tiger's stripes draw the eye towards the outline of the surrounding rocks and branches. Because of its highly adaptive coloring, the cat can move to within feet of unwary prey. This is why the tiger is often called a phantom by forest peoples; it seems to appear out of nowhere.

Speed and strength are also essential. A 400-pound cat can spring 20 feet and take down prey many times its size. Once the tiger grasps its prey with sharp claws, struggle is futile, as the cat quickly severs the neck with its fangs. Larger animals are choked to death by clamping down on the throat, and exceptionally large prey, like a 2,000 pound gaur, require the tiger to grab the back and flip the creature onto its neck in order to snap it. Despite the tiger's strength and explosive attack, only about 10 percent of attempts are successful. The element of surprise is everything.

Unlike lions, which hunt in large prides, tigers almost always hunt alone. Lions encircle and run down their prey—an adaptation to the open savannah. The tiger creeps and ambushes—an adaptation to the forest. So when a tiger kills a big animal like a boar, he has to keep his hard-earned meal safe from other predators and scavengers by hiding it in a secluded area, usually a cave or wooded enclosure. Again the dense forest provides this protection. A tiger can feed on a carcass for several days, especially enjoying the soft putrid flesh. Here again the tiger is different from the lion. Because game is scarcer and the tiger hunts alone, he must be as diligent in keeping his kill as he is in hunting it down.

The tiger is the supreme predator wherever it lives. And this role is essential for the health of the entire forest ecosystem. Biologists refer to tigers as "keystone" species because they keep an entire ecosystem intact—like a keystone in an archway. The tiger is also called an "umbrella" species because it stands atop the entire food chain, ensuring stability at lower levels—much like a king over its forest subjects.

The path now took us along the open meadows that flanked the lake. The dense green forest that covered the hilly terrain offset the blue water and blue sky. The hilltops were naked, covered only by tall grasslands, the result of the constant stream of cool air in these higher elevations. I spotted a herd of sambar deer, *Rucervus unicolor*, similar to the American elk, only smaller, across the lake. And high on the grassy hilltops, a dozen or more gaur were peacefully grazing.

As we rounded our final bend, we reached the lake and I could see the boat dock in the distance. It was next to the old colonial hotel, now converted for tourists. Dozens of people lined up for short wildlife tours on old wooden vessels that plied about the lake. We steered clear of the noise and commotion and entered the main road a couple hundred yards away. Despite the carnival-like atmosphere, I took the obligatory boat tour the next day.

◆ ◆ ◆

Periyar Lake is dotted by hardened gray skeletons of half-submerged trees that died a century ago, and it features countless coves and finger-like channels formed by the partial submersion of the rolling landscape. Boats can be hired to ply the lake to view deer, gaur, and elephants. Birds and yard-long mahseer—silvery, more streamlined cousins of the carp—are common. But the wary tiger rarely makes its presence known.

After an hour of drifting along the shore and spotting deer, elephants, bison, and monkeys, I asked the captain if he ever saw tigers. He said no, that they usually stayed in the forest or came out at night. "But a few months ago, they did find a dead tiger floating in the water on the restricted side of the lake," he said.

I asked him if he thought it was the result of foul play, poison or something. "I don't know," he replied. "Most of the illegal killing here is for elephants and their ivory." He pointed to a herd that was bathing in the shallow water. It had no tuskers.

Like the tiger, the Asian elephant (*Elephas maximus*) is a forest animal. And unlike its African cousins, female Asian elephants have no tusks, and males are either with or without them. He shook his head in dismay as he explained that one of the favorite ways poachers kill elephants is through electrocution. He pointed to some high-voltage wires that ran over the hills surrounding the lake. "Poachers will cut these wires and lay them along elephant paths. When the animals touch them, they're killed instantly. The poachers hack out the valuable tusks with machetes or gouge them out with chain saws, leaving behind the carcasses to rot."

Most ivory winds up in East Asia. And just like tigers, elephants are fighting to survive throughout Asia. Millions once lived in the forests stretching from India to China. Today about 40,000 remain. Over half are in India, with 1,000 found here at Periyar, which represents one of the highest populations in the world. In the future, tuskers may become rare or even extinct. As a result, the elephant population may stabilize at Periyar after all the tuskers have been poached, as there is little demand for tuskless elephants.

Before leaving Periyar, I visited one other place where tigers are known to prowl. That was the old hunting machan (tree stand) that sits on a forested peninsula across the lake. To get there I rode in a small motorboat whose ancient engine struggled for a good half hour to deliver me to the landing area, a stone staircase that emerged from the water and ascended the steep shore leading to an old hunting lodge, which once belonged to the Maharaja of Travancore. It is now a luxury hotel. The machan was several hundred yards behind the lodge.

I had been warned by the boatsman that a young male elephant had been seen on the peninsula. As I moved away from the hotel grounds and onto the dirt path, I saw elephant dung scattered everywhere, some of it steamy fresh.

My heart beat a little harder as I moved deeper into the forest. I reached a wide grassy opening, but the eight-foot grasses blocked my view of everything on either side of the path—good tiger and elephant habitat. I stayed alert.

Elephants are actually the more dangerous animals, killing more people in India every year than tigers. Rogue males and females with calves are the most aggressive. They run people down and tear them apart with their trunks and then stamp the remains into the ground in an unrecognizable pulp. I had heard that one way to divert an angry elephant charge was to drop articles of clothing behind as you ran for your life. The elephant stops and tears up each piece before proceeding. With luck, enough time can be gained to escape. I pulled some clothes from my backpack and clenched them in my hand as I picked up my pace.

Finally, I breathed a sigh of relief as I rounded a bend and spotted the machan. It was tucked into a clump of woods overlooking a meadow that rolled down toward the lake's edge. The old structure stood 10 feet off the ground like a giant tree house. A deep moat surrounded it to keep out wild animals, namely elephants.

The maharaja and his entourage reportedly shot many animals from this elevated perch. But when he decided to create a wildlife preserve around the lake in the 1930s, hunting came to a sudden end. The machan lay dormant for decades until it was added to the tourist venue in recent years. Despite its tourist use, it has received little care—probably not since the last royal hunt over 70 years ago.

The inside was small, about 12 feet square, with hard wooden benches built against the walls for sitting and sleeping. Sadly, almost every inch was covered with graffiti left behind by Western tourists. One actually gave an account of a tiger sighting in the meadow below. The rest was banal and vulgar. What had once been the exclusive domain of Indian royals was generously opened up to the public and then promptly desecrated.

I spent the rest of the afternoon exploring the surroundings. I followed a path into the heart of the forest but only saw assorted rainforest birds and giant Malabar squirrels. I returned to the machan as evening set. During this time of day, tigers often leave the forest for the thick grasses along the water's edge, to ambush animals that congregate there to feed and drink.

I perched myself in front of the window that overlooked the meadow and lake. A few birds called out as the sky darkened, soon replaced by a chorus of crickets. Rats started to scamper about on the roof. One ran into the room and stared at me for a second and then darted back into a hole in the wall. I heard a soft rustling outside. When I looked back out the window opening, I saw a hundred sambar grazing in the meadow. Their eyes shimmered orange in the moonlight as they glanced my way. For hours, as I tried to stay awake, I observed them as they continued to graze.

When I awoke the next morning, the meadow was vacant. The only creatures were a few cormorants in the distance on the lake's shore. Birds and squirrels had been active since dawn, but they were quieting down as the day grew hotter. Bees and wasps hovered around the flowering shrubs surrounding the base of the machan, and agama lizards sprawled out on the path, taking advantage of the morning sun.

I looked out of the window a while longer in the hopes of glimpsing some wild creature. But as mid-morning approached, I knew my chances had grown thin, as the heat would soon stifle all animal activity until evening. Then, all of a sudden, I caught movement on the far shore—about a half mile away. A mass of yellow fur moved along the edge of the forest. I pulled out my binoculars and got a look as it stopped. It was a tiger! I watched it for a few minutes as it strolled in the morning sunshine and then rounded the bend, out of view.

I have seen many wild animals, but none like this. All those tiger images—cunning, grace, beauty, danger—rushed through my head. They grew in intensity as I stood looking into the nearby woods, sunlight and shadow, like black and yellow stripes, the quiet rustling of leaves like a secret prowling cat. Maybe there was another nearby. I looked deeper into the dimly lit woods, fixated as a cold sweat over came me. Surely this is the king of the jungle.

I made my way back to the boat dock, walking along the forest path, watching the tall grass on either side, thinking now of tigers, not elephants. I paused, stopping to peer through grassy openings, hoping to see another cat. I did not, and the boatsman eventually came and fetched me. But the exhilaration stayed with me the rest of the day.

In many ways Periyar is a model tiger reserve. Grassland habitat was expanded through the creation of the lake, allowing for more prey species and more tigers. Abundant and visible wildlife brought tourists, who contribute much-needed income to the area. There are, of course, tradeoffs, as thousands of acres of forest have been permanently flooded, and hundreds of thousands of people visit the park every year. But this is often the only tradeoff available in this crowded, poor land. The project is also active in developing economic alternatives for local people so they put less pressure on the forest and tigers.

During my last days at Periyar, I met up with a man named Vijay, a young man just out of a graduate program in anthropology, who heads up these programs. He did not look like a typical forestry officer as he wore traditional clothing rather than the standard military-style forester's uniform. As we sat down and talked, he explained that the casual dress was part of Periyar's new approach to building greater trust with villagers. And then he told me that his work at the project was part of his personal philosophy as a committed naturalist.

"I live a simple life and don't own many things. I walk or bike everywhere, and only take the office motorcycle to visit the far side of the sanctuary," he said. "I must live the part of an environmentalist, as well as enforce the laws!"

Then we talked about the project and I asked him about the priorities. The goal was to develop local economies that would reduce the pressures on the forest—namely farming, timber felling, and hunting. He worked mainly with tribal villages scattered about the park boundaries. And a few villages still exist inside the park.

These tribal people are the original inhabitants of India, having entered the subcontinent from east and west tens of thousands of years ago. Like the tiger, they were forest dwellers who lived by hunting, as well as gathering and simple agriculture. Today India has about 80 million tribal peoples sharing virtually every forest with tigers across the land. Because of their intimate knowledge of forests and animal life, these people were used as hunting guides and trackers in the past. Today many use these skills illegally, to poach animals and timber. A goal at Periyar is to create economic alternatives to reduce these crimes. One program recruits and pays tribal men to run anti-poaching patrols. Another pays them to run jungle safaris. Both have had success.

Another project goal is reducing fuelwood demand. Forest people rely almost exclusively on wood as their cooking fuel, and almost all comes from the forest. But rather than relying only on punitive measures to control harvesting, the government is also trying to reduce overall demand for wood. One way is to provide more efficient cooking stoves, some of which can cut wood use by half. Vijay explained that a new stove was designed at a local university. It is of simple brick construction and any villager can build one. "You just have to have all the right angles and things like that to trap more heat," he explained.

He had his motorbike with him at the office that day, so we got it and drove out to a nearby village where Vijay showed me one of the new stoves. It was built in back of one of the huts, where we were soon joined by a large flock of children who gathered around as we inspected the stove. A large stack of wood, about four feet tall and just as wide, lay off to the side. Vijay pointed at it. "You see this big pile; it will last only a few days. Imagine that stack multiplied by thousands of fireplaces all around this reserve. This puts tremendous pressure on the forest. Fortunately, these piles are less now," he added cautiously.

People seemed to like the new stoves, but like many policies in India, implementing is a slow process. It is also hard to change people's habits. Many still cook with the old inefficient stoves or open fireplaces as their grandparents did. "It will take some time to convert everybody." He added with a hopeful smile, "But for these children the new stoves will be normal. They are our hope here at Periyar."

He then brought me around to the back of the village to show off another innovation that will help reduce pressures on forests—improved agriculture. Agriculture must begin to change. It must become more intensive, more efficient, and more productive. Like everywhere in India, the people of Kerala have been farmers for millennia. Because of the demographic pressures in the state of Kerala, for centuries, people have been growing spices, vegetables, and fruit in tiny plots to maximize production. Agronomists and ecologists are now building upon these practices to create "permaculture," or permanent agricultural systems. The idea is to pattern agriculture after the surrounding forest by using intensive planting methods with many types of plants, better plot designs, and without expensive chemicals. And if agriculture is more efficient, then more land can be set aside as forest, and this means more wildlife, including tigers.

I asked Vijay about these types of methods. He was not familiar with the technical term permaculture, but he explained that more efficient farming was a project goal. We visited a plot on a steep hillside that had never been intensely farmed. Now with the project support (especially for seed and saplings) villagers grew an intense garden of fruit trees, vegetables, and staples like manioc and cassava. Goats and cattle grazed, keeping down weeds, and chickens roamed about eating insects. It was a wonderfully integrated and productive system, all on a few acres of land.

We strolled down a small path that took us through a grove of trees. I asked him if he had seen any tigers in the two years he had worked here. "Yes, a few months ago I saw one in a forest clearing, and people see them while out trekking around the lake. One morning I found footprints of a tigress with cubs right along the main reserve road. So they are around." Vijay beamed as we turned back towards his motorcycle.

The tiger is a wildly fecund creature, as single tigresses can produce 30 offspring in a lifetime. And if the cat is given enough forest, it can sustain its population. So there is always excitement when cubs are sighted because it is a sign of a healthy tiger population and hope for the animal's future. But the tiger needs forests to breed, to hunt, and to roam. It needs places where it can remain "king of the jungle." And only people can make this happen.

Vijay dropped me at my lodging a few minutes later and we exchanged goodbyes. He turned and waved and drove off to the tiny bungalow on the outskirts of the park he calls home.

—2—

The Art of Tiger Hunting

It is hard to believe that the tiger, now threatened everywhere, was once routinely hunted. And nowhere were tigers hunted more fervently than India. India was synonymous with tiger hunting. It had many tigers, whose perfect forest habitat—open woodlands and grasslands—also made tracking and shooting easier. India also had some of the world's biggest and most beautiful cats. As a result, countless thousands were killed over the centuries. Many were killed for sport or as trophies, others for their valuable bones and other body parts, which had many social and economic uses. Tigers were also killed out of self-defense or exterminated because they were considered cattle-eating vermin. Some were even killed for food.

The hunt, or sometimes slaughter, occurred throughout India and took many forms. Tigers were shot with bows and arrows and guns. They were poisoned and caught in steel traps and driven into pits and giant nets, where they were speared or clubbed to death. They were pursued with horses and dogs or on foot. And they were shot from the safety of machans and the backs of elephants.

Tigers were also killed in large numbers especially by the rich and powerful. Hindu maharajas, Muslim nabobs, and other nobles had a great passion for hunting, or shikar, as it is known in Hindi. And when the British came to India they continued this tradition, establishing such "field sports" as a favorite pastime.

Despite this carnage, tigers were so common that their overall survival was never placed in jeopardy except on a localized level. At the eve of World War II, tens of thousands still roamed the subcontinent. Hunting only became problematic for the tiger when combined with the drastic loss of forest habitat. More importantly, the relationship between hunter and hunted changed dramatically with the spread of modern communication, transportation, and weaponry.

By the latter half of the twentieth century, the cat had lost all of its advantages and defenses as hunting became increasingly rational and efficient.

The outcome was predictable—the cat could easily be tracked and killed. As the balance in hunting tilted decisively in favor of the hunter, it led to widespread slaughter, causing the Indian government to finally ban tiger hunting in the late 1960s, when the cat population dropped below 2,000.

The tiger is now at the complete mercy of human beings, and tiger hunting is now unthinkable, its very mention almost sacrilegious. Yet, to dismiss tiger hunting as pure ignorance and bloodlust is inaccurate. It is more complex and must be examined and understood in the broader context of the cat's historical relationships with people.

◆ ◆ ◆

I visited an old tiger hunting ground, the Bandipur forests that lie just north of Periyar in the tri-state region, where Kerala meets the states of Tamil Nadu and Karnataka. This area is one of India's best wildlife habitats, with about a thousand square miles of contiguous forest interspersed with another few thousand square miles of farm and grazing land. It covers an area of the Western Ghats known as the Nilgiri Hills, whose range of elevations contains many forest types.

The United Nations dubbed the region the Nilgiri Biosphere Reserve, a designation for particularly rich ecosystems. It comprises Bandipur Tiger Reserve and several other sanctuaries that protect a few hundred tigers, one of the largest concentrations in India. Deer and boar are common, and the reserve is a major elephant and primate sanctuary as well as home to one of the largest remaining populations of the rare Indian wild dog, or dhole.

As I approached Bandipur, I saw a few signs for small tourist lodges along the roadside. I spotted one that looked inviting, and my driver promptly veered our jeep off the highway down a small dirt road that led into the woods and then a clearing, where a few bungalows stood on 10 acres of land. They were simple one- or two-room wooden buildings with slightly slanted roofs and verandas in front. The setting was not striking, but pleasant—there was a tall mountain behind me and good views of the surrounding scrub forest.

At the main compound, a group of Indians sat drinking afternoon tea, including the owner, a man named James. After a brief introduction he led me to the bungalow where I would stay for the next several days. James was in his forties and dressed in Western attire; he spoke good English, acquired by time spent in England. He was back in India for good, he explained, and had invested in this lodge to accommodate the growing number of tourists who visit Bandipur and other sanctuaries each year.

As for tiger sightings, he told me he had had the fortune of seeing one recently, pointing in the direction of Bandipur. "I was on the village bus and a tigress and her cubs were sprawled out, sleeping under a tree by the roadside. The bus stopped, and we all looked on in amazement. The cat finally woke up from all the commotion. But she was not afraid. She just got up and slowly strolled back into the forest." He described several other encounters in the last few years, all along the forested road to Bandipur.

Though what he told me next I had not expected, it was more than a pleasant surprise. Not only was this lodge an old colonial forest retreat, now used by government officials for business and pleasure, but it was where the famed English hunter and writer Kenneth Anderson stayed when visiting this area. James pointed to a small faded building down the hill. "He slept right there in that one. This is where you will stay," he said. As we walked toward it James explained it was the only building out here decades ago. "Nothing but wilderness back then—lots of tigers and other wild animals."

I put up my things and sat on the front porch. A lone hoopoe, a brown and yellow bird with a curious fanned crest, pranced about on a distant rock as the sun moved towards the horizon. Otherwise all was still.

It was an odd feeling staying in the very place occupied by one of the most famous tiger hunters ever. I studied the surrounding forest and wondered how many big cats had been shot out there. Had they been skinned in this very compound? I glanced over to a clump of old trees. Maybe their striped pelts were stretched, cured, and dried only a few hundred feet from my front porch. The thought that never seemed far from my mind returned as I gazed back at the forest: how many tigers were still left out there today?

The first tiger hunters were tribal peoples. They shared the forest with tigers, competing with them for prey and territory. The relationship was often violent, but also equal. Men slaughtered the great cat, and it retaliated in kind. People accepted this tradeoff uncritically, because both people and tigers were part of the great cycle of life to which death and suffering was natural.

These early hunts were risky and required great skill, courage, and luck. The hunter had to mimic the tiger, stalking silently and waiting patiently, often for days, until the cat was found. For the modern hunter the appearance of the prey signals the climax—when the animal is finally shot and killed. In the past it was only the beginning. Once the cat was spotted, the hunters, who almost always moved in groups to keep the odds in their favor, fastened their eyes upon their target and trained their spears or cocked their bows. To get into range they had to get close to the cat, so every move could give them away. The tiger could attack or retreat.

Hands and nerves had to be steady as the moment of attack arrived. The aim had to be true, as any misfire quickly turned the hunter into the hunted. As soon as the cat was hit it roared in anguish, sending shudders through the forest and down the spines of men. A wounded cat could shift the entire momentum in its favor. Flailing paws and snapping jaws severed limbs and heads and sliced bodies to shreds.

Hunting on foot was dangerous business, which is why many forest tribes preferred catching tigers with snares or large nets constructed of twine or other natural fiber. The tribesmen unfurled the nets, stretching them for hundreds of feet through the forest. Others would drive the cat towards the nets. Still other men then closed the circle and tightened the net until the tiger became tangled in it. They then clubbed or speared the unfortunate cat to death.

Tribal hunts were largely practical affairs. Tigers were killed for self-defense or to rid an area of a competitor for wild game; they were killed for their pelts, talons, and canine teeth, all used as ornamentation and to bring protective powers to those who wore them. Tigers were also slain for valuable body parts used for medicine and magic. Gall supposedly cured bone diseases, fat eased rheumatism, and powdered bone, a host of ailments. Whiskers, eyebrows, dried skin, and even the animal's tail purportedly drove away evil spirits and brought good luck. And in some parts of India, especially the eastern Himalayas near Myanmar, tiger flesh was consumed for strength and courage.

Despite widespread hunting and use of the animal's parts among the forest tribes, tiger hunting was always embedded in fear and reverence. Strict rules and codes determined the hunt; tigers were killed in specific ways, in specific places, and for specific purposes. Taboos frequently governed the use of the animal's parts, and acts of repentances and forgiveness followed a successful hunt. After dividing a tiger's body the head was often saved. It was placed in a stream or propped in the crotch of a tree, with mouth open so the tiger spirits could escape and not seek revenge against the hunters.

There are many tribal communities scattered about Bandipur. One morning a guide, who also served as a driver, took me by jeep to one of the villages located at the base of a mountain several miles away. Occasionally we stopped to take in the scenery of rolling hills covered in a dense carpet of green, a tangled mass that looked inhospitable to humans but perfectly inviting to wild creatures. In the past, this type of landscape covered hundreds of thousands of acres throughout southern India. We were looking at one of the few remaining swaths, now part of a national forest.

He halted when I spotted a gaur scampering up a steep incline in the near distance. It stopped briefly to look back and then bolted into the forest cover. The gaur (*Bos gaurus*) is basically an oversized wild forest cow, with a taut black or brown coat and white legs that look as if they were painted on. Despite its huge size, sometimes reaching 2,000 pounds, the animal is shy and only becomes dangerous when cornered or threatened. Gaur are fairly common in the Nilgiris, and tigers here are particularly adept at hunting them.

We soon approached our village, where maybe 150 people lived. We first spent time with the villagers, my guide translating as I asked questions about tigers. "They are around here," my guide stated. "But they never cause problems."

Then a group of us wandered along the periphery, gazing into the forest for signs of life. We spotted a flock of forest babblers in a nearby thicket. Then an older man gestured to the hills above to a small herd of elephants. They were half a mile away and hard to track as they moved among the trees and boulders that were the same color and shape as the elephants. "Elephants are very, very dangerous," my guide exclaimed. The elder then spoke to him. My guide's eyes lit up and he shook his head. "The old man told me one had killed a woman recently as she was out gathering fuelwood. It was a young male. They are most dangerous!" I looked back at the old man, who was staring down at the ground, shaking his head in disbelief.

It is hard to protect rural people from rogue animals—whether they are elephants, leopards, bears, or tigers—as shooting them is now illegal. In the past, villagers organized themselves and killed these rogues. Hunting a wild animal such as a tiger was an expression of self-defense. It was also a part of these people's freedom as they dealt with the world around them on their terms. But with the rise of governmental power these primeval acts, as well as other traditional freedoms, have been taken away from forest peoples.

As we prepared to leave the village I asked my driver what kind of work people do here now. Some farm (shifting cultivation) and herd livestock—practices adopted from the lowland farmers. Women are still permitted to gather wood and herbs from the forest. Some collect lichens, which have natural dyes, and sell them to paint makers in the city. There is even a special project coordinated by a local nonprofit group that helps the women market their goods.

A group of young men sat under a shade tree by the aqueduct. They waved and nodded politely as we drove off. My driver told me that the village men sometimes work for the forest department doing various odd jobs. When I asked whether these jobs helped the tribal communities, he laughed: "They could, but these people don't like to work. They are lazy. All they want to do is roam around in the forest. They will never change."

This attitude towards tribal people was common among farm and herding folk (like my guide). It is an ancient antagonism that has yet to be resolved.

The first great changes in tiger hunting occurred with the rise of farming and herding (settled agriculture), which began to replace hunting and gathering some 10,000 years ago. These cultures developed superior technologies and more complex social and political organizations.

Tiger hunting became easier and more intense as forests were cleared, weapons perfected, and animals domesticated. Hunting also became a profession. Tribal peoples and specific castes protected villages from predators and hunted game for income. Hunting also became more ritualized. The growth and concentration of the farming population created a huge pool of laborers who could be used to drive cats towards waiting hunters, perched safely in treetop machans or on elephant back. The use of trained elephants greatly shifted the odds in favor of the hunters, allowing them a safe location for shooting and better means of tracking cats over large areas.

Elephants were used primarily in the tall grasslands of the Himalayan foothills, or the *terai*, which stretches across northern India (including what is

Nepal and Bhutan today). Machans were used in drier areas of west and central India, where elephants were not native, or in densely forested and mountainous areas that prohibited their movement. Dry weather was usually chosen for hunting, when the vegetation had passed its explosive monsoon growth and began to wilt or shed its leaves, making tracking and shooting easier.

Not only did the hunt change in size and scale, it took on a new expression—that of the inordinate social and political power of the upper castes in the rigid Hindu social system. For the wealthiest princes, shikar (the hunt) was expanded to become a spectacular ritual known as the *hunquah*, involving every sector of society. When a tiger hunt was organized by a nobleman, word went out to the surrounding villages. An army of servants and other retainers, often numbering in the hundreds, was organized.

The most dangerous jobs were those of the "beaters" and "stops." These men spread out across the forest to form a huge triangle, with the stops at the sides and the beaters the base. The beaters drove the tigers out of the forest by shouting, blowing horns, and beating pots (hence the term "beater"). The stops, more skilled and knowledgeable about tiger behavior, were posted in trees or on rocks at intervals to watch the cats' movement and to make sure they did not break out of the triangle. When a cat approached the boundary, the stops, like the beaters, shouted or yelled to scare the cat back into the triangle and toward the hunters. The hunting entourage formed the apex toward which the tigers were driven and then shot.

Despite reduced danger for the hunters, these ritualized hunts had their risks and tigers still had a chance. Primitive weapons—spears and arrows—often missed their target, and cumbersome muzzle-loading guns misfired, allowing the tiger to escape. Sometimes several cats broke out of the forest at once, creating pandemonium. Once the tigers knew they were trapped, they attacked with abandon. A crafty cat could turn back on the beaters to break out of the line, maiming or killing men in its flight for freedom. Others hid in brush and grass only to be forced out into the open gunfire.

Elephants were trained for courage, using their size and bulk to deflect the tiger's assault, and the elephant driver or mahout had the most important task of keeping the pachyderms steady as the cats either tried to escape or attack. But if a tiger latched onto the elephant's trunk, clawing and biting into the soft tissue, the huge animal could panic and stampede, often giving the cat its freedom. The hunting party could be thrown to the ground, where they were at the mercy of tiger attacks or the elephant's massive feet. Angry tigers even scaled the elephant's side, dragging men to their death.

The ritualized hunt reached its pinnacle under the Mogul emperors, Muslim conquerors who ruled India after the sixteenth century. The hunt became more than simply a quest for tigers and other game. It became one of the main expressions of Mogul political power, a veritable military expedition that involved thousands of men and beasts that went forth into the forests for months at a time. Hunting was used to train soldiers and elephants for battle.

The long forays were also used to survey the periphery of the empire—to gather information on local peoples and their rulers as well as the surrounding forests and terrain.

The Moguls perfected the "ringing in method," which they called the *qumargah*. The French physician to the Mogul court, François Bernier, gave an account of one such operation near Delhi, in which 200,000 men took to the field and hunted for months in an area that covered several hundred square miles. Sentries guarded every entry point with the utmost vigilance, and anyone who trespassed or hunted illegally was severely punished. Thousands of animals large and small, including tigers, were taken using a range of weapons and hunting methods.[1]

The ritualized tiger hunt was a paradox. It was purely egoistic, a display of martial skills and bravado. It was also a brazen display of social power that aristocrats had over people and wild animals. But at another level, the hunt was an intimate part of the human struggle with nature, expressed in ritual and symbol. The tiger was the supreme expression of nature's danger and mystery. Thus killing the cat represented a human victory over these primordial forces. Even a simple peasant understood this symbolism. And after a tiger hunt had ended and the corpses of the striped victims were paraded through town and village, local people exalted their lord's victory.

Ironically, the aristocratic desire to kill tigers also helped save them. Virtually every tiger reserve and wildlife sanctuary in India today was once a private royal forest. Thus, without the foresight of the royal hunter—albeit for selfish reasons—there would be far fewer tigers in India today. Ironically, the people who had killed more tigers than any group in history ensured their long-term survival. It was another example of the intimate connection between tiger and lord.

Bandipur Tiger Reserve was the hunting ground of the rulers of the kingdom of Mysore, which was centered in the adjacent plains. It was one of India's greatest princely states. Flat fertile land and a rich supply of timber and elephants in the nearby hills made it a military and economic power unmatched in southern India.

Tigers were so common here and so much a part of life that one of Mysore's rulers, the notorious Tipu Sultan, fashioned himself as "The Tiger of Mysore." He was so enamored by the cats that his entire palace became a tiger shrine. He had a gold-plated tiger throne and wore tiger-striped clothes. His soldiers wore striped uniforms, had striped guns, and carried a banner proclaiming, "The Tiger Is God" (despite the fact he was a Muslim). He kept dozens as pets and fed his war captives, especially the loathed British, to them. Eventually the British defeated him in 1799, and Tipu died in battle, fighting like a tiger with his striped sword in hand.

Bandipur remains one of India's better tiger havens because of the dry open woodlands and abundant streams. Large grassy clearings are common and favorite haunts of spotted deer that congregate there in huge herds. On my first trip into the park I counted over a hundred in a single grouping and saw many more throughout the day. I did not see any tigers, but the abundant deer no doubt supported a healthy cat population, which the government estimates at over 50.

On another excursion, I drove to the border of the park and stopped at a grassy knoll that spread down a hill towards a stream that was shaded by bamboo and tall trees. This would have been a perfect setting for a tiger hunt. I could envision the great drama unfolding—long lines of men filing into the forest and fanning out to create their trap. Then the yelling, beating, and blowing of horns began a mile or more in the distance, growing in intensity as the beaters marched steadily forward. The nobles, adorned in ornate hunting attire, were perched atop their elephants that created an impenetrable line in the forest clearing. The hunters remained silent, eyes fixed on the dense forest ahead, waiting for the cats to break cover.

And then came the great culmination as the tigers broke into the open. Beaters yelled and screamed, elephants trumpeted wildly, and the trapped tigers roared with rage. Then muskets sounded, one after another, followed by more yelling and commotion, the final anguished roars of the cat as they died, followed by silence.

It was a well-scripted drama that captured the eternal struggle between tigers and men. It endured for millennia, unchanged. But, like the tribal hunt, it too slipped into history, never to be revived.

Langur monkeys clambered about on some nearby trees and brought me back into the present. I stayed a little longer in the hopes a tiger might appear. The midday sun was hot, with hardly a breeze for relief. The heat and the monotonous saw of insects put me in a haze. Then I saw movement to the far side. Unfortunately, it was only a man, a villager with his herd of cows. He moved slowly as they grazed, completely oblivious to the government restrictions on trespassing on wildlife sanctuaries.

In the past these low-caste herders were considered criminals for trespassing on royal land, just like tribal hunters, who were deemed poachers. They were punished by violence or even death. Today the Indian government owns the forest and its tigers, and although penalties for infringing on these rights are less than in the past, the struggle for control continues, another act in this age-old drama between ruler and ruled.

On my return to the lodge I stopped to watch some mahouts wash their elephants at the river. It was a natural pool, perhaps 10 feet deep and three times as wide, formed by a large outpouring of lava that spread across the river like a dam. The elephants stood in the shallow end as the men scrubbed them with large brushes. The elephants assisted by dousing themselves with water, soaking animal and trainer alike. When finished, the mahout mounted the elephant and strode into the deep water, where they swam awhile, and then returned to the shore where all gathered in the sun to dry.

These were all captive-bred animals. But in the past, wild elephants were caught and then trained throughout India. Bandipur was one of the most famous areas for elephant capture before the British ended the practice in the late nineteenth century as wild populations dwindled. But for centuries prior, elephants were herded into large enclosures called keddahs, similar to tiger nets, but much sturdier. Once they were encircled, their captors carefully roped and secured their legs to avoid injury. They fed the animals a steady diet of their favorite foods, sugarcane and bamboo shoots, which slowly pacified them.

But sometimes angry elephants stampeded, breaking down the keddah and killing people and each other. Like tiger hunting, it was a risky affair.

Once pacified, elephants would be selected and trained by mahouts for their work potential. The strongest and more courageous were used for fighting and hunting, and the remainder became beasts of burden. Today fighting and hunting are no more. Elephants like these now work for the forest department or take tourists on short stints through the jungle in search of their old nemesis, the tiger —though now on peaceful terms.

Although Indian royalty hunted tigers for millennia, it was the British who became the most enthusiastic tiger hunters of all time, shooting more tigers in a shorter period of time than any single group of people in history. And it was the "Great White Hunter" who epitomized both hunting and empire. He was usually a military man, sometimes a wealthy aristocrat, who spent his life roaming the far-flung colonies for the thrill of the chase. The tiger captured his imagination even more than the great beasts of Africa, which is witnessed by the hundreds of books and articles written exclusively about tiger hunting. Through their prolific writings, these men introduced the tiger to the imagination of the average European, presenting the animal as something to conquer and kill—one of the many exotic prizes of empire.

When the British came to India in the early 1600s as traders and merchants, they quickly developed a keen appreciation for shikar. Officers and administrators readily adopted the ancient hunting rituals, taught to them by Muslim and Hindu rulers who eventually became their political subjects. While the British enjoyed the pomp and ritual, they also used shikar, much like their predecessors, to cement relations with their new subjects and to survey territory.

It was not until the late 1700s that the British East India Company, a commercial entity, began to use British military might to control larger parts of the subcontinent as the Mogul Empire slid into decline. However, after the Sepoy Mutiny by native soldiers in 1857, the company proved its inability to administer such a huge population and territory. The British government took over India as a crown colony—the British Raj.

The Raj rapidly changed India and tiger hunting, as the British integrated the subcontinent into the growing industrial economy. Roads and rails spread rapidly to facilitate trade, move people, and extract natural resources. This gave hunters unprecedented access to once-remote tiger forests. Hunters were also equipped with new and better weapons, breechloaders and later cordite rifles, which increased killing efficiency. A tiger could now be dropped with a single bullet from a great distance.

The hunting ethos also changed. While "shooting," as the British called it, was still largely a privileged sport, the British also believed in democracy and extended hunting rights to the common subject employed in the Raj. Many soldiers and civil servants took advantage of this once-elite privilege to "bag a tiger" before their India tours ended. In a way it was reward for duty in a harsh and lonely land, where disease, discomfort, and death were part of life. It was also believed that hunting upheld certain standards. Field sports, like military drills, kept men in physical and mental shape and also kept them from drinking and womanizing, pastimes thought to enervate rather than invigorate.

The Raj began the golden age of tiger hunting, at least from the hunter's vantage. In the 100 years between 1840 and 1940, British big game hunters shot some 20,000 tigers—over 10 times the number surviving in India today! Tigers were also eradicated as vermin. And when these numbers are added in, the total number of tigers killed is probably two or three times that figure.

The British colonial government paid bounties to professional hunters and villagers alike as tiger eradication was viewed as a sign of social progress. The naturalist and hunter G. P. Sanderson, who actually spent 13 years in the Bandipur forests, once claimed the mass extermination of tigers one of the most exciting and glorious sports afforded to man. He poisoned 90 cats during his career.

Many Britons shot a century (100 cats), and some killed even more. Montague Gerard killed over 200, and a Colonel Nightingale shot 300. George Yule of the Bengal Civil Service shot 400 in 25 years, before he stopped counting, although he continued to hunt. The writer and sportsman Gordon Cumming shot 73 cats in two years, and William Rice shot 93 in four summer vacations. Almost every high-ranking British official and visiting dignitary had tiger shikar on his or her itinerary.

Traditional Indian rulers were not to be outdone by these parvenus. Stripped of their political power and social prestige, many retreated to their forest kingdoms, turning to shikar to display the only freedom and power still allowed them. The Maharaja of Scindia killed about 700 during his lifetime. And the Sultan of Surguja claimed 1,700—probably more tigers than any person in human history.[2]

Yet, as human power over tigers and nature grew, a conservation movement also emerged in British India. By the late nineteenth century, the British became more prudent as they recognized that tigers served important ecological functions. They controlled deer and wild pigs that destroyed crops. Too few tigers led to increased wild game populations, which then plundered farmers' fields. Too many tigers reduced wild game populations, which turned tigers into cattle lifters and man-eaters. This realization created a more prudent approach to hunting. The British saw their role as the balancers between tigers, men, and prey animals.

The slaughter of wildlife also changed hunting attitudes and policies. British hunters had already decimated the Asian lion, cheetah, and rhinoceros populations, forcing hunting bans on these animals by the early twentieth century. The elephant had already received protection in 1870. Tigers were still abundant, and as forest creatures they were able to escape the onslaught that befell other Indian wildlife that lived in the open plains or northern grasslands. Still, tiger shooting became more restricted, especially as warnings emerged about the cats' potential demise, even in areas where they had once been common.

Big game hunters issued many of these warnings. In one of his famous books, *Man-Eaters and Jungle Killers* (1957), Kenneth Anderson predicted the problems that would beset the tiger and India's wilderness in the coming decades:

> I know localities where until 1930 the moaning sough of a tiger or the guttural sawing of the panther were normal sounds in the night. . . . Where once the pug-marks of a tiger and other wild-animal trails would tell their morning story of the creatures that had passed that way during the night, the tiny tracks of a few rabbits might today indicate that they at least have not been exterminated.
>
> . . . Anyone who has never come to know and love the jungle, its solitude and all that its denizens signify, could never appreciate such sentiments, nor the sense of irreparable loss and sorrow felt by those who look for the once familiar forms that are no longer there, or listen vainly for those once familiar sounds that were music to their ears, only to be greeted by a devastating silence.[3]

Kenneth Anderson was born in India and lived there his entire life, staying after its independence, when most of his countrymen left for good. He was a civil servant by trade who worked in Bangalore, but hunting and nature were his passions. He enjoyed shooting as sport and was never involved in the massive ritual hunts. He was a practical hunter, a "tiger slayer"—a job given by the colonial government to kill man-eaters and other rogue animals that threatened rural peasants. And he almost always hunted alone, usually staying in local villages or a forest bungalow like the one I was renting. His preferred method was sitting up in a machan over tied bait—a live cow or goat—which attracted the tiger. Sometimes Anderson even had to stand vigil over a human corpse, whose body had to remain in place so the man-eater would return.

Few today can imagine the terror man-eating tigers or leopards instilled in isolated rural villages. In the past, a single man-eater could kill hundreds of people, and man-eating sometimes became so bad that entire regions were abandoned. Economies collapsed, leading to starvation, disease, and famine. Man-eaters were among the craftiest cats and required great skill and patience to bring them to justice. They were aggressive, which made hunting them particularly dangerous.

One of the most notorious killers Anderson encountered back in the 1930s was the Hosdurga man-eater. The tiger killed several people just north of Bandipur, including a young Englishman, a novice hunter, who naïvely thought he could stalk and kill the wily cat on his own. His half-eaten body was found a day later.

Anderson eventually tracked the cat down after many months. He describes his final frightful encounter:

> To my left rose the snarling head of a tiger, its ears laid back in preparation for the spring, its jaws wide open to reveal the gleaming canines.
>
> My bullet crashed into the wide-open mouth, as the animal launched itself forward. Rushing blindly on to gain higher ground, I all but lost my eyes in the intervening lantana, the thorns tearing my flesh and clothing, while pandemonium broke loose behind me.
>
> With the back of its head blown out, that tiger tried to get me, and when I had covered those remaining fifteen yards and spun around, it was but two yards away, with a great gaping red hole where half its skull had once been. Almost beside myself with terror, I crashed a second, third and fourth bullet into the beast and as, shattered, it toppled on its side, I sat on a piece of the ruin, shaken, sick and faint.[4]

Despite his many blood-chilling tales, Anderson was never as famous as Jim Corbett, who became world renowned after writing about his tiger hunting exploits in the Himalayas in the earlier part of the twentieth century. But Anderson was the more pivotal figure in the history of tiger hunting. He lived through that age of carnage following World War II, when the old skills and rituals of the tiger hunt had given way to mass slaughter.

Indian independence in 1947 changed everything. It was a time of exultation because of freedom from British rule, but also one of turmoil as society changed from aristocracy to democracy. Industry, cities, and population grew dramatically. Pressures on the tiger increased as forests were cleared and agriculture expanded. The hunting and conservation ethos also changed.

The new Indian government soon monopolized hunting, just as the British Raj, Moguls, and maharajas once did. Unfortunately, game laws and rules were also widely disregarded during this period, often for lack of trained personnel to uphold them, but also as acts of purposeful violation.

The members of the new bureaucratic ruling class wanted to enjoy and display their new political privileges. And what better way to express this in both practice and symbol than to shoot a tiger on a former maharaja's private reserve. Many also wanted to make money by commercializing shikar. Rather than it remaining a privilege given to the few, shikar outfitters brought clients from all over the world. Many were wealthy Americans eager to join the ranks of the maharajas and the British as tiger hunters.

With international air travel and affordable shikar packages, tiger hunting proceeded with a frenzy. The tiger hunting craze continued through the 1960s. Thousands of pelts left India in that decade alone. The tiger population plummeted from an estimated 10,000 cats in 1950 to less than 2,000 in 1968.

For men like Anderson, who had spent their lives in the Indian forests and had been active sportsmen, it had to be a melancholy end to their lives to witness the tiger's demise. Fortunately, Anderson did live to see the government's ban on tiger hunting in 1968. He died in 1974, a year after Project Tiger, the comprehensive government program to protect India's remaining tigers, was enacted.

◆ ◆ ◆

My last day at Bandipur was memorable as I spent it climbing the mountain behind the lodge. It was tall and equally steep and was strewn with enormous volcanic boulders. The forest was dense, shading a trail that cut deep inside it and wound its way up and over the steep hill before ending in a collection of towns and villages in the neighboring valley.

James had secured a guide for me the previous day, and we were off early the next morning, as the trip to the valley and back would take an entire day. We first entered a bamboo thicket that covered the base of the mountain, where a large pool had formed. It is common to see several clumps grow together, creating extensive shade forests where wild animals are known to gather. Elephants are particularly fond of the many succulent shoots found there.

The bamboos here in southern India are immense, with four-inch stalks that tower 50 feet high and grow in stands that are just as wide. There are several species, all of which have a unique life cycle. Every few decades they will burst into flower and then die off en masse a few weeks later. The entire countryside is then littered with massive stacks of collapsing dead bamboo. But with the rains, tiny bamboo shoots will sprout up around them, grow to maturity, and repeat this spectacular process—on cue—many years later.

As we approached this large grove we spotted elephant tracks and piles of dung. Once again I was on guard to their presence. My guide told me that an elephant in a nearby forest had chased him and several other men several months back. They actually ripped off their shirts as decoys that allowed them to run to safety. Fortunately we saw none that day.

The only potential danger we encountered was a small viper that slithered a few inches from the trail farther up the mountain. Later we came upon animal droppings. My guide thought it was a tiger, but I was not so sure—maybe a leopard, or even a jackal. There were no tracks to confirm our findings. As we climbed higher we had a commanding view of the forest and the vast Deccan Plateau in the distance, an unending patchwork of farms and villages.

I walked out on a large boulder that jutted high over the forest and took out my field binoculars to scour the surrounding hillsides. It was steep terrain, with every inch covered by giant boulders and rubble, and trees of varying sizes grew up among them. I could see a few streambeds, or nallahs, filled with smaller stones where the monsoon rains poured into the lowlands each summer, but they were now dry. Higher up I spotted a cave on the side of the mountain.

I fastened upon it for a long time, hoping some creature, maybe a tiger, would emerge. But none did.

The forest was tranquil, except for the sounds of the occasional bird and the low constant hum of insects. A gentle breeze brought various forest fragrances with it. This was the world Kenneth Anderson had loved. Like most hunters, he enjoyed the hunt—the ornamentation and tactics, the pursuit and the thrill of the kill. But there was always more to it. It was the total experience—the long days spent in the Indian jungle, the sights, smells, and sounds, and being close to wild animals—that was cherished most.

Modern readers judge these men, especially the Great White Hunter, harshly and perhaps deservedly. For they represented wanton killing, elitism and paternalism, and many other vices no longer acceptable or fashionable. Yet I wondered as I sat out in the middle of this forest whether these criticisms are not also tainted with a hint of envy. These men still lived in a world filled with natural beauty and danger, where tigers were so common they were killed without remorse. It is a world that every modern naturalist wishes still existed and one that these hunters never thought could end. It has, and with it came an end to the art of hunting the greatest trophy animal of all time.

—3—

The Land of a Thousand Tigers

Poaching is currently the greatest threat to the Indian tiger. Some 1,500 cats have been killed across the country since the early 1990s to supply tiger parts, mainly bone, for traditional Asian medicines. The greatest demand for tiger parts is from China; the greatest supply of tigers is India. The cat is caught in this violent economic intersection. As the supply of tigers decreases, price goes up. As the demand for tigers increases, prices go up further. Tiger bone is now in such short supply and great demand that its value per ounce has surpassed that of many precious metals.

Tiger poaching is nothing new. The poor have always broken laws to kill problem tigers. So have the rich, to get that trophy at any cost. But in the past these events were isolated and localized. Modern poaching is different in degree and kind. Poaching has evolved into a distinctly modern form of tiger hunting in which money and technology are the driving forces. As a result, tigers are no longer killed out of defense or survival or even as trophies, but as commodities harvested, traded, processed, and sold on global markets.

In the past the number of tigers was so great and the demand relatively low that tigers were never seriously threatened. China once had tens of thousands of cats—as late as the 1960s. But the Chinese leader Mao Zedong ended that. He killed off the country's tiger population in a few short decades as part of his "Great Leap Forward," a program designed to rapidly modernize the "backward" countryside by clearing anything that stood in the way of progress—peasant villages, forests, swamps, and vermin like tigers. It was a tremendous success. China transformed its countryside, modernized its economy, and reduced its entire tiger population to a few dozen animals.

Mao's ambitious project killed thousands of tigers, creating stockpiles of bone and other tiger parts that supplied traditional medical practitioners for decades. The stockpiles finally ran out in the 1990s. And that was when new supplies were needed. They were found in India and Southeast Asia. And the demand for these tigers is now being fueled by a richer, more populous and powerful China. They want tigers and their demands are being met.

Local crime syndicates, as aggressive and cunning as any multinational business, organize the actual killing and transportation of tigers. They utilize modern communications and mass transportation networks, and they kill with state-of-the-art weapons, traps, and poisons.

Their lucrative incentives can corrupt villagers and government officials alike, presenting grave dilemmas for the Indian government. How can an already corrupt and inefficient government bureaucracy be kept from being seduced by dynamic and powerful poaching networks? And how can policies and programs convince poor rural people that they should protect the tiger rather than kill it? What kinds of policies are most effective in stopping poaching?

Many believe that government must invest more in manpower and technology—train better personnel and buy better equipment—jeeps, telecommunications, and weapons. Most Indian forest guards are armed with simple walking sticks, while their criminal counterparts often have AK-47s. Others argue that it is the villagers who must be won over first. The government must invest in economic development, provide alternatives to poaching as well as other types of forest exploitation—create incentives to protect rather than poach, create alternative economies.

Both are important ways of stopping poaching. But, they are not enough. Poaching is at root a political problem, a question of will and leadership.

First there is the problem of political priorities. The Indian government can build a nuclear arsenal, a transcontinental rail system, a globally competitive high technology industry, and massive hydroelectric dams (that have effectively flooded thousands of acres of wilderness), but it cannot protect its tigers. The reason is simple—conservation interferes with the drive for economic development. The desire to become an "advanced industrial democracy" is the reigning religion of Indian leaders (the only Western religion to succeed in India). In their eyes, conservation does not create jobs, does not add to the GNP and income, and it cannot be taxed. Ecology gets in the way of economics.

Then there are ineptitude and corruption. The two feed off each other, creating incentives to commit crimes—a market to be exploited. Some of the worst problems are at the judicial level, where lawyers (not surprisingly) and judges are bought off by powerful crime syndicates. A labyrinth of Indian laws and procedures that slows the legal process to a virtual standstill thwarts even honest efforts at prosecution. There are currently hundreds of wildlife cases pending in New Delhi, and some take a decade before coming to court. Others never do. The defendants skip bail, move, or die. The overall conviction rate for wildlife crimes is below 5 percent.

The case of Sansar Chand is a telling example. Chand is one of the most successful poachers in India, largely due to his ability to manipulate the system. He is responsible for the death of dozens, if not hundreds, of tigers and other wild animals across the land; he has been caught many times but has always avoided jail time. He usually posts bail and then flees. He has also bribed policemen and judges or hired sharp lawyers. On one occasion he was caught then faked a heart attack, began to cry, and avoided being jailed. He later escaped.

Chand finally received a lengthy prison term in 2005. He then admitted the extent of his operations. It is a family business, one that has thrived for over 20 years. Despite his widespread operations, Chand explained he is only one of 20 poachers in India, all with a dozen or more operatives in the field. So one poacher was caught and jailed, but sadly for the tiger there are many more, ready and eager to take Sansar Chand's place.

Sansar Chand's most infamous poaching operation was in the state of Rajasthan in western India. It was particularly destructive because it was an inside job, involving high-ranking wildlife department officials who provided easy access to the vulnerable cats. Dozens were killed in a few months. Today fewer than 50 tigers remain in all of Rajasthan, a state once known as "the land of a thousand tigers."[1]

◆ ◆ ◆

Rajasthan is a dry state, mostly desert. Tigers only live in the far eastern corner, where the last monsoon rains are captured by the towering Aravalli Mountains to support scrubby, savannah-like forests. It is the driest tiger habitat on earth.

In addition to tigers and majestic desert scenery, Rajasthan is the historic home of the Rajputs, fierce warrior clans known for their martial and hunting skills and a penchant for ritual suicide—widow burning and mass military suicide campaigns. However, their violence was offset by art and architecture that is unrivalled anywhere in India.

There are two tiger sanctuaries in Rajasthan—Ranthambhore and Sariska—both once Rajput strongholds with a history of violence where tiger poaching does not seem entirely out of character. Ranthambhore is the more famous, a bellwether of sorts, its history foreshadowing the fate of tigers everywhere in India.

It was a royal hunting ground and battleground for centuries. The sanctuary is named after a thousand-year-old fortress that still stands inside the park. In colonial times Ranthambhore was a gathering place for dignitaries. King George of Greece, the Duke and Duchess of Gloucester, Princess Zia (daughter of Czar Nicholas the Second), Woolworth heiress Barbara Hutton, Lord and Lady Jersey, Lord Mountbatten, and other famous names can still be found on the old shikar register.

One of the most famous tiger hunts at Ranthambhore was in 1961, when Queen Elizabeth II and her husband, the Duke of Edinburgh, shot two tigers on their honeymoon. But the old thrill of the hunt was quickly undone by public outcries back home. They were unexpected, but they made an impact. When the couple arrived in Nepal a few weeks later, Prince Philip did not hunt again, claiming an infected trigger finger. The Ranthambhore tigers were the last shot by British royalty, an event that signaled the beginning of a popular conservation ethic in both Britain and India.

In 1973 Ranthambhore became one of the first Project Tiger reserves. Its natural beauty and abundant tigers soon made it the most photographed, filmed, and written about place in India, even on earth. President Bill Clinton visited India in 1996 and he too went to Ranthambhore. But by then the poaching crisis was unfolding.

I arrived at Ranthambhore after several days of arduous bus and train rides, settling in at a small hotel, one of the many that lined the main road into the park. It was a simple concrete structure run by a family that lived next door. I was anxious to visit the park and made arrangements for an excursion the next day. Ranthambhore is one of the few places in India where wild animals are frequently seen, because of the park's small size, open forests, and good network of dirt roads. The number of park visitors is also limited, and tours are scheduled for certain times and follow set trails, reducing congestion and the impact on wild animals.

The weather was cool and the sky clear blue the next morning when I met my guide. After a brief introduction we drove off to the park, only a few miles away. The park boundary was easy to spot—even from the distance—a distinct line of woods contrasting sharply with the barren village lands. As we came closer we passed a sturdy stone wall erected on the sanctuary side to keep wildlife from wreaking havoc in the villages, or vice versa.

The park entry was a giant archway carved of stone, once serving as the sentry post to the fortress inside. It was wedged into a narrow opening of a sheer cliff that towered hundreds of feet above us, the flat tops of which were roosts and nesting places for dozens of vultures. We entered and descended into a streambed surrounded by surprisingly large leafy trees—much different from the surrounding scrub. Ghost trees (*Sterculia urens*), named for their bright white bark, grew up on the hillsides. They become particularly conspicuous in the dry season, standing out like giant bleached skeletons against the grays and browns of the dormant forest. Their bark almost glows at night, which made them convenient markers for nighttime travelers years ago. They were now in full foliage, but their stark white bark could still be seen from far away.

Then we came to a giant grove of banyan trees (*Ficus benghalensis*), covering at least an acre. It was actually a single tree that reached this enormous size by sprouting aerial roots from its branches that grow to the ground to form new trees. The result is a procession of limbs, roots, and stalks from the main trunk, thus earning the banyan the name "walking tree."

Because of its size (up to 4 acres) and age (800 years), the banyan is an important Hindu icon, often the site of shrines or temples. The labyrinth of branches also provides refuge for forest creatures, including tigers, which have been known to frequent the one right in front of us. It is a perfect hideout, the light limbs and dark shadows mimicking the tiger's striped form. It is also a good place to ambush animals that gather at the large pool in the adjoining nallah.

We continued along the sandy road that wound through the flat bottomland, stopping and waiting for tigers whenever we came to clearings. We moved like this for a good hour and saw monkeys, sambar deer, and birds, who were heard but not seen in the dense vegetation. The "may-haw" of the peacock frequently pierced the morning air, but we saw no tigers. We drove off to the hills, in hopes our luck would change.

The higher elevations gave us good views of rolling grassy hills studded with small, dark-leaved trees. The landscape was reminiscent of west Texas or California—but with tigers. The hills spread for miles, broken up by deep ravines and escarpments and finally tall, barren mountains that formed a huge wall around us in the distance.

We stopped at a place that overlooked a valley with a small stream, crowded in by dense stands of grass—good tiger habitat. So I began to search with my binoculars and saw movement right away. My driver also saw it, and we both pointed and then turned to one another with eyes wide open. A tiger? We quickly looked again. Maybe another chance. But, it was gone.

Then there was more movement, straight ahead. Five small shapes walked slowly across a field, pausing to graze. Perhaps deer. I trained my binoculars again and saw they were gazelle—the Indian gazelle (*Gazella bennettii*) or chinkara. They are not very striking, lacking the impressive horns of their African counterparts. But this species does have some status, as it is closely related to the mountain gazelle (*Gazella gazella*) of the Middle East, which is the namesake and parent species to the entire tribe.

We drove a while longer until we came to a place where we saw the great Ranthambhore fortress off in the distance. It is a masterful construction, with ornate walls and ramparts stretching up and across the rims of towering cliffs, thus integrating the natural features of the mountain into the overall defensive scheme. It was virtually impenetrable for centuries, but it eventually fell to the Sultan of Delhi in the 1300s. After a protracted siege, the surrounded and outnumbered Rajputs finally opted for *jauhar*—ritual suicide—rather than surrender. The Rajput leader, Hammir Deva, and his soldiers donned their saffron garments and met their opponents in a fight to the death. Meanwhile his wife built a huge pyre inside the fortress, in which she and the entire village burned themselves alive. The fortress fell, but no Rajput endured the humiliation of defeat.

Ranthambhore is now silent, yet it maintains an air of invincibility, having withstood the heat and the creeping forest for centuries. But a millennium takes its toll, even on such a massive structure. The walls are slowly crumbling, shrubs have colonized open ground, and vines weave over and through the network of rooms and alleyways. Troupes of macaques now man its turrets and copulas, acting as modern-day sentries, always alert to the presence of the tigers who enjoy the many hideouts below.

Pictures of tigers skulking about these ruins is what made Ranthambhore world famous. They have almost become the icons of Indian wildlife photography, in which nature and culture are always joined together, very different from African or North American wildlife photography, which emphasizes pristine wilderness—nature untouched by humans. But such places are almost nonexistent in India. The two are always intertwined and have been even before the days of Hammir Deva.

We were soon joined by a carload of tourists, busy chatting and laughing and alternately swooning at the scenery, oblivious to the carnage that is intimately tied to this beautiful land. These modern Indians, like their Western counterparts, want to see the old ruins, the stunning landscape, and, of course, the great striped cat. But they have no interest, or perhaps stomach, for the violence that is so often a part of beauty. They are romantics.

But the local villagers are different. The screams of warriors on their way to their final battle, the stench of burning carcasses, the blood of thousands spilled onto the rocky ground are embedded in memory. These events are relived through ritual and prayer at the shrines and temples built throughout these forests. Many are built in places of great suffering or tragedy, which like places of great beauty, push the human imagination and spirit to their limits.

It was approaching midday and my tour was soon to end. We drove back to the main gate when we suddenly stopped, and my driver looked down on the road. "Tiger tracks," he exclaimed. The deep imprints in the soft sand extended for about a hundred yards before disappearing into the forest. My guide thought he had been here this morning, about the same time we were driving on the other side of the park. Hit or miss!

We drove up to where the tracks disappeared, and I got out of the jeep and bent down to grab the pugmark (footprint), letting the sand run through my open fingers. So close! My guide smiled and shook his head in sympathy. He then glanced back to the woods nervously and gestured that I return to the safety of the jeep as the cat could be nearby. We sat for a while longer quietly watching the road, hoping he would return.

We were just ready to leave when a striped form sauntered across the road and stopped 20 yards ahead and looked directly at us. "It is a young male. I have seen him before," whispered my guide. He just stood there and looked at us and then turned, with nonchalance so characteristic of all cats, and entered the woods. We started our engine and went to the spot, hoping to catch another glimpse. We caught his backside as he melted into the shadowy woods.

That evening I visited a retired senior forester named Mr. Singh to discuss poaching. He was there in the beginning, when Ranthambhore first became part of Project Tiger. He helped the tiger population grow and made the place famous. The hotel owner had mentioned him to me, so I asked if we could drive out to see him. He lived right outside the park, in a small bungalow in the forest close the main road.

A servant greeted us as we parked by the house and promptly went inside to give word of our arrival. The door opened and we entered the main room, where Mr. Singh stood. He was not tall but imposing, dressed in field green with a full head of white hair and a beard to match. He had a broad-rimmed hat and walking stick by his chair. His skin was weathered by the desert heat, and his eyes were penetrating, sharpened after years observing the forest, following the tiger and learning its secret ways. He was steeped in the old ways, mannerly and respectful, yet aloof. After all, he was a Rajput, one of a long line of rulers who, like the tiger, had presided over this land for centuries.

We went to his study, filled with tiger pictures and other mementoes collected during his many years working at the park. He sat us down and then explained his struggles at Ranthambhore. He battled with local villagers who did not like his many regulations and the government who often refused to enforce them.

"It was difficult in the beginning. We had to set up a whole new system of rules that told people they could no longer go to the park. They had grazed the cattle and collected firewood there for centuries. Naturally there was often conflict. It was very heated for a while, tensions were high. People resorted to violence. Forest guards were threatened, and some were even murdered. But we prevailed in the end," he said with confidence.

I then asked him about poaching at Ranthambhore. It peaked in the 1990s, when over 20 cats were killed within a few years' time. For a moment I could detect a hint of anger. He sighed and then explained how the situation worsened after he retired.

"The staff became greedy and corrupt. There was no discipline. That is why poachers killed the tigers. The staff was involved. Fortunately the new manager is a good man and things have improved greatly. Today the tiger population has rebounded. I hope it will stay that way. It all depends on the personnel. They must be devoted to the tiger and its protection. That, and that alone, is what will save the cat."

Then we discussed the tiger's future. He was pessimistic. "There is not enough space for them in this crowded land, so tigers turn to cattle and man-eating. That is why people kill them. Even the Sunderbans [the mangrove forests in the east], which many think has hundreds of tigers, does not. There are man-eaters there for a reason. There is nothing to eat, so the few remaining desperate tigers eat people." Then I asked about poaching. "You see what happened here. Poaching is happening everywhere in India. The tiger is in real trouble." He did not elaborate further, and I did not press him as it was getting late. The conversation soon wound down and we readied ourselves to leave.

As we returned to the hotel, the main drag, its shops and gaudy signs, the rows of diesel-belching cars and jeeps all came back into view. Despite the low-level chicanery and the ineffable tackiness that accompanies the tourist trade, not to mention the noise and garbage, it may have saved Ranthambhore's tigers precisely because people come here and spend money to see them. Even the government has recognized that money talks.

We pulled up to our shabby little hotel and both walked to the front door. I shook my friend's hand and thanked him for taking me to see the old forester. I told him that he has a fine little hotel and that he and his family have done a great service to Ranthambhore and its tigers. He smiled and then bobbed his head to and fro, as only an Indian can, and thanked me in return.

I went on many more park excursions during my stay, but I never saw another tiger. It was frustrating and disappointing as the cats were certainly around. I ran into guides and tourists who had all seen them. I too had close calls and come across many more pugmarks. But the cats were always gone before I arrived.

The greatest obstacle was the forest itself, still green and lush after the summer monsoon; it makes it hard to spot wild animals. The abundant water also disperses wildlife throughout the park. In the dry season they are forced to congregate along shrinking streams and waterholes. This is the best time to see the cats. This is also when the tiger hunters of past and poachers of present find them.

But for the naturalist, these fresh forests and fields have their own singular magic. In the mornings deer and antelope, monkeys and songbirds are active. Reptiles emerge to sun themselves in the afternoon—mugger crocodiles on the shores of large pools and desert monitors, five feet of muscle and sinew, along the stream banks, where they search for prey. Big black vultures, their naked heads barely visible, soared a mile overhead.

And then there were all the human artifacts—the ruins, shrines and temples, and the network of footpaths. Some were tucked away in obscure places, abandoned and overgrown, but all telling some story about this forest, its people, and its tigers.

Sadly, too many visitors come here and miss the forest for the tigers. They look for tigers as if it were a relay race or treasure hunt. And whenever I encountered visitors, the question was always "Did you see a tiger?" It is to be expected at a tiger sanctuary but becomes exasperating. On one occasion, I sarcastically (but politely) responded to one inquirer that I had not seen a tiger, but I did see a lion. To which he told me I was very lucky as they are quite rare now. Indeed.

The tourists were getting to me and more were arriving. So it was time to leave. My next destination was Sariska, which lies just a few hours north by train.

Sariska is even more impressive than Ranthambhore—a landscape of mesas and flinty cliffs overlooking forested valleys and small lakes. And there are also more cultural artifacts, reflecting the turbulent history of the region—from the Aryan civilization dating back 5,000 years to the Buddhist period 2,500 years later and the Gupta civilization, 1,000 years ago. Most lie in ruin, destroyed by furious battles, earthquakes, weather, and the work of vandals and art dealers.

Sariska was once the private hunting reserve of the Maharaja of Alwar, Jai Singh Prabhakar Bahadur. He was a controversial figure who was removed from power by the British because of incompetence in 1933. The maharaja was exiled to Europe, never allowed to return to India, and died in a Paris hospital a few years later. The British claimed Jai Singh was mentally unbalanced and prone to excessive cruelty. Supposedly he once doused a polo pony with kerosene and burned it after losing a polo match. He was also said to be a sadist who snatched young women and children off the streets and raped, tortured, and even murdered them.

The maharaja's supporters claimed otherwise. They said that the maharaja was a proud man, a Rajput who did not accept British rule. So he defied them. That is why the British banished him. He once ordered a fleet of Rolls Royces from Britain and then used them as garbage trucks throughout his kingdom to insult his colonial superiors. The maharaja was also a keen sportsman who often outperformed his British superiors on hunting expeditions.

Admirers also remembered him as a prudent conservationist who only allowed a handful of tigers to be shot at Sariska every year. As a result, it always had many tigers during the maharaja's day. But this changed rapidly after independence when the reserve, only a hundred miles south of New Delhi, became a favorite shooting ground for the new government's ruling class. It became one of the worst cases of abuse in the country, as tigers and other game were shot without rules or remorse. By the late 1960s only a handful of cats survived.

Sariska's tigers recovered in the 1980s like those of Ranthambhore. But the poachers, also conveniently based in New Delhi, have consistently targeted the sanctuary since that time. Their attacks have been devastating. And many now wonder if the tiger can survive here and whether it is worth the effort to try combat this unrelenting criminal onslaught.

One of the maharaja's most enduring legacies is his hunting lodge, actually a palace, whose massive columns and gleaming white exterior create a dramatic presence in the middle of the forest. The lodge, now a luxury tourist hotel, is fully restored, complete with swimming pool, exercise spa, and tennis courts. It has dozens of guest rooms and huge halls outfitted in oriental finery and an entryway unchanged from the past, complete with hunting trophies, including a stuffed tiger measuring 11 feet. The old shooting tower still stands behind the lodge, its heavy iron ring, where "baits" were tied to attract tigers, now rusted into the ground.

The lodge was my first stop at Sariska. But luxury is not my idea of fun, especially in the middle of a wildlife sanctuary, so my taxi driver took me to the more modest (and much cheaper) government hotel, located in a pleasant grove of woods but unfortunately too close to the noisy Jaipur highway. It was also crowded, being the only other accommodation at Sariska.

Sariska, like Ranthambhore, has many dirt roads covering much of the sanctuary. Tigers and other wildlife are most commonly seen in the main valley, known as Kalighati or "dark valley," named for its thick forests where light barely reaches the forest floor. The forests are not tall, even around the nallahs, but simply dense and shady. The valley is also dark because of the preponderance of olive-leaved *dhok* trees (*Anogeissus pendula*), a type of acacia. Splashes of bright green salar trees occasionally break the monotony, but acacias are dominant.

Like all of western India, Sariska has many acacia varieties, all important to local people, mainly as fuelwood. Dhok is especially hard, making it burn slow and hot. But the most highly prized acacia is catechu, because of its reddish aromatic wood used to make the slightly intoxicating stimulant and breath freshener known as *paan*. It is used all over India, especially by men, which explains the constant chewing, drooling, and spitting, as well as the red-stained teeth acquired after many years of use. It is prepared from wood chips, boiled for hours into a paste, then sun-dried into a powder and combined with other spices to create the popular product.

The great demand for paan and fuelwood makes these valuable trees constant targets for timber poachers who sneak into the park from neighboring villages. Unlike tiger poaching, which is organized by large crime syndicates, the illegal timber trade at Sariska is localized, with many participants seeking diffuse targets, making it almost impossible to control.

I toured the Sariska forests many times during my stay. Peacocks and langur monkeys were the only creatures spotted with any frequency. Even the common sambar deer was usually heard rather than seen. Its warning call, a loud "dhank," is one of the most familiar sounds of the Indian jungle, alerting other forest creatures to danger. Human observers, if lucky, and led by a skilled guide, can follow the ruckus to find a tiger or leopard.

The act of listening and tracking is the essence of the Indian safari. In Africa, large animals are present in such abundance, and in open areas, they are seen with ease. But, in India, where animals are fewer and forests thicker, locating them is more difficult. In many cases the only way to find a big cat is to follow forest sounds, and the sambar's call is one of the most reliable signs that one is nearby. However, sometimes the jungle sounds can be confusing, as the wily tiger has been known to imitate the sambar call to lure the animal towards him! This behavior is called "pooking," and although some naturalists have disputed it, many hunters and trackers of the past have observed it.

One afternoon I hired a guide and drove into the forest and just waited and listened. I was lulled by the low hum of insects and the periodic singing of birds, when a sharp "dhank" sounded about 20 yards away. My guide looked at me and raised his finger and whispered, "Tiger!" Another deer sounded off, followed by the angry chatter of langur monkeys in the treetops. More deer and more monkeys joined in, creating a crescendo of animal sounds that moved slowly towards us, following some animal below. We dared not move, hoping the creature would emerge on the road in front of us. The furious chatter increased, then a peacock joined in with its wild "may-haw." The entire forest was whipped into a frenzy.

Then all of a sudden, a ball of yellow and black bolted across the road and disappeared again into the woods, triggering the sounds of more monkeys. For a moment I thought it a tiger, but it was too small. It was a leopard. Not the great striped one, but nonetheless thrilling. The cat kept moving, and we tracked it by following the noise of the forest. It eventually made it to the rocky hills that overlooked the valley, and the deer and peacocks and monkeys, no doubt satisfied with their vigilance, finally quieted down.

Leopards (*P. p. fusca*), or panthers as they are locally known, are still relatively common in India. In fact, worldwide they are the most numerous of all the big cats. Some estimates say there are 500,000 leopards—over 100 times that of the tiger.[2] It is also the most widespread big cat, found throughout Africa and in pockets across the Middle East and Asia. Its success is due to its craftiness and smaller size, allowing it to live off birds and small mammals, although it can kill larger prey. In India its range overlaps with the tiger's, but the leopard prefers the peripheral forests, often living close to villages where it commonly preys on livestock. It is also especially fond of village dogs; these are habits that make leopards widely hated.

But the fortunes of the leopard may be changing in India, as it is now being killed to provide a substitute for tiger bone. My guide told me that Sariska has about 50 leopards, with more in surrounding forests. He then told me about one known to frequent an abandoned fortress—Kankwari Fort—in the neighboring valley. So we went off to see it, hopefully to encounter another leopard, but also to view—in good romantic fashion—the magnificent ruins.

The fort sits atop a lone hillock in the middle of a wide valley—strategically situated to halt any thoroughfare. It is a classic castle, built of huge blocks of dark brown granite stacked a hundred feet high and topped by square merlons where archers once stood. Large cylindrical copulas stand at every corner of the rectangular structure, giving views of the entire valley. The Rajputs built it a thousand years ago, and then it sat abandoned for centuries until the Mogul ruler Aurangezeb used it in the 1600s to imprison his main political rival, who also happened to be his brother. The poor soul spent many years in captivity before finally being executed. The fort's importance declined after the British took power, when it was again abandoned, and it has stood vacant ever since.

Kankwari Valley is also the historic home of the nomadic Gujar peoples.[3] A few dozen families still live in small, square earthen huts at the foot of the fortress, as they have for centuries. They are pastoralists, herding cattle and growing only dates. Their entire income is derived from the sale of milk and dates to surrounding villages. A jeep visits them weekly to bring the Gujars and their wares to local markets for sale. The cash earned is used to purchase food and other staples.

We parked our jeep at the village, and my guide got out to talk to an elder sitting in the shady doorway of his hut. He asked if any leopards or tigers had been near the fort recently. The old man shook his head. He then started talking in a loud voice, making frantic gestures with his hands and pointing to the main road we had just driven in on. My guide told me that there is a big tiger who has been prowling about this area for months. We thanked our man for the information and then walked the remaining hundred yards to the fort.

A narrow road led to the fort's single entrance, once covered by a heavy wooden gate, now long gone. We moved cautiously up and went inside, just in case some wily cat made this his lair. My guide shouted and clapped his hands as a precaution. There was no response except from a far copula, where a white scavenger vulture (*Neophron percnopterus*), also known as the "pharaoh's chicken," sat, its dingy-white feathers blowing up over its naked pink head. It moved slowly then gave us an evil eye before plunging into flight, leaving behind a trail of feathers.

Grasses and shrubs covered the fort's interior, and mature trees shaded over once-open courtyards. The walls of interior rooms were painted with beautiful frescos of flowers and other ornate designs. Many were faded and some had fallen victim to youthful games of target practice by local village boys. My guide explained that these Muslim artworks have no value for the Hindu Gujars. Some are even purposely destroyed for that very reason. I walked over to examine the paintings closer, so painstakingly created to satisfy aristocratic tastes, now scattered on the ground, covered by mounds of guano from the bats and birds who now lay claim to this place.

A narrow staircase went up to a copula that gave views of the entire valley. All the land surrounding the fortress and village—hundreds of acres—was barren, because of centuries of grazing and wood cutting. Dead stumps were everywhere, mixed in with date palms, whose frail forms only accentuated the starkness of the treeless plain. Denser palm groves were clustered about the villages and the small lake at the base of the fortress. The village, the fort, the water, and the feathery palms created a stereotypical oriental setting worthy of the Daniell brothers, the British painters who made their fame painting these "exotic" images over 200 years ago. It was certainly beautiful, serene, and sublime, but not the best place for tigers.

The sight in front of me captured the whole struggle between people and tigers in India. This small village of a couple hundred people had effectively denuded several hundred acres of prime tiger forest, a process going on everywhere in the country.

I gazed back out at the forest and heard a faint sound in the distance, which I traced to a herding boy. He was singing while moving his flock of cattle through the forest. It was a quaint picture, but it also reinforced the constant presence and disturbances of people, even these simple Gujars, on the tiger's forests.

It was now late afternoon and we made our way back before the park closed, taking an alternate route to see another part of the sanctuary. My guide had already spent many more hours with me than was required, and this route would take a while longer. So I tipped him for his extra time. He thanked me and told me that this is not work to him. He enjoys being in the forest as much as possible and loves the tiger.

"I helped count them on the most recent population survey," he said. "I sat up in tree stands for hours looking across the open land for tigers while the park rangers went into the forest to count pugmarks. There are twenty tigers at Sariska," he announced proudly. I asked if the future looked good. "Yes I think so," he said confidently.

We ascended the rutted road out of Kankwari Valley, passing another herding boy and his many cattle on the way. He waved and then returned to watching his precious herd. A few minutes later we rounded a bend and immediately saw a huge black animal lying in the middle of the road. It remained motionless—obviously dead—as our noisy jeep approached. "It is a buffalo!" my friend announced. "A village buffalo," he repeated.

We pulled right up to it and instantly saw the huge red gash in its neck. The head, swollen in the midday heat, was almost completely severed from the body. Flies swarmed around the wound, and a large puddle of blood, now soaked into the dirt, spread all around the carcass.

I began to look around for more evidence and instantly saw the pugmarks of the killer. It was a big tiger. Hoof marks and paw prints were mixed up in a confusing pattern, a dance of death in which the cat had apparently latched on to the buffalo's throat, whirled it to the ground, and suffocated and killed it.

"A male tiger," my guide said. "At least ten feet long. He is probably the one the old man told us about." My guide shook his head in disbelief. "The villagers won't like this."

This was not the ordinary emaciated Hindu sacred cow, but the more valuable buffalo, a large work animal, also a store of wealth and sign of status.

The government is supposed to compensate villagers when wild animals kill their livestock. But it usually takes months to get the money. And then it is often too little. Sometimes excuses are made and the people are never paid. So the villagers just get fed up and retaliate, poisoning carcasses to kill the perpetrator. The poachers also take advantage of this situation. Disgruntled villagers can easily be bribed to kill off problem tigers.

As we drove off my driver explained that this is the problem when people and tigers are so close together. There is conflict.

"This is why the government wants to resettle villages like Kankwari."

Then he explained that many villages have been resettled over the last decade.

"Kalighati once had many residents, but they moved willingly because they were given good new land outside the park. But there is less good land now because there are three lakhs [300,000] people here. It is even worse for these herders. They need open land for their cattle. Only the park has open land now. So they want to stay."

"Will the villagers try to kill this big tiger for killing their buffalo?" I asked.

"They will be angry. But I think the government will do the right thing. The poachers here killed so many tigers a few years back. They want the villagers to be happy. It is a slow process. But I think we will work it out. Sariska is the best tiger sanctuary, even better than Ranthambhore. Don't you think?"

"I hope so." I sighed.

We drove a short distance and then came to a hill where the dark velvety forests of Kalighati sprawled out in front of us. We paused a moment to take in the view and then drove down into them and headed back.

After 2004, park visitors began to report fewer tiger sightings. Suspicion grew that something was amiss when tourists were taken on "wild tiger chases" in which no tigers, even pugmarks, were ever seen. After initial denial, the Indian government finally admitted that poachers had killed off Sariska's last 15 to 20 cats.

Sansar Chand, in his last criminal spree before being caught, was responsible. He had been killing Sariska's tigers successfully for 15 years, but he had never managed to wipe them out entirely. A serious attack came in the early 1990s, around the same time that Ranthambhore was hit, and over half the cats were killed off. But the tiger recovered and the population increased again to over two dozen. But Sansar Chand persevered, working the system until it finally paid off. The only good news from the Sariska fiasco is that this stunt eventually led to his arrest. But for the tiger it came a few years too late.

Biologists reintroduced three tigers in 2008 (caught at Ranthambhore), with the hopes of adding three more in a few months' time. But the fear of future attacks at this small and vulnerable sanctuary raises many questions about the wisdom of this policy. It will take years for the population to reestablish itself, and many believe the project is doomed to fail. Kalighati has never been so dark as it is today.

—4—

Project Tiger and the Legacy of Jim Corbett

Many critics blame India's poaching crisis on Project Tiger itself, the government program created in 1973 to save the cat. The project established tiger reserves in India's various habitats, each receiving special government funding and personnel. The original 9 reserves have now grown to 40,[1] covering 15,000 square miles. But the tiger population has plummeted from over 4,000 in the 1980s[2] to 1,500 today.

Furious debates now rage in Indian conservation circles over the project and its future. Many believe the problems are organizational. The project is too top heavy, with staff and resources concentrated in New Delhi, rather than in the field where they are needed. Some argue the project should be more decentralized, devolved to the various state governments in the Indian federal system, or should include nongovernmental and international organizations and local communities. Some believe there are simply too many projects, causing scarce resources to be spread thin. Fewer reserves with more resources could protect more tigers.

Others see government as the problem. Project Tiger has followed the fate of all bureaucracies—constant expansion. The project, like the bloated Indian state it is part of, squanders funds on buildings and salaries, meetings and travel, and reams of official documents and glossy brochures. Corruption is inevitable.

Some critics, like the naturalist Valmik Thapar, go even further and claim that the government is fully aware of the tiger crisis, but rather than address the problem it creates new reserves to give the appearance of successful conservation. In reality, most are "paper reserves," existing only on maps and in law, but not in practice. Thapar believes the government not only creates fake reserves but fake tiger numbers to match. The government purposefully inflates tiger numbers to protect jobs and the status of Project Tiger.

These charges have brought into question the time-honored practice of counting "pugmarks," or footprints, in the government tiger census, held every four years. It requires experts, who supposedly can differentiate tigers based on their individual pugmarks, to scour the entire country for footprints and then come up with a grand total of tigers. Many believe this is not only a huge waste of funds but wildly inaccurate as well. It is also privileged information that can easily be manipulated.

All these problems are real, and something needs fixing. But perhaps what Project Tiger needs most of all is a leader who can make tigers and conservation a political issue. Project Tiger needs someone like Indira Gandhi, the prime minister who started Project Tiger almost four decades ago. She faced a scenario similar to the current one. Tiger numbers had plummeted and government policy was ineffectual. She struggled against long odds to save the cat, but she had a vision and the will to prevail. When the project finally become law, it was hailed as one the greatest political and conservation victories ever, setting a conservation precedent for all nations—rich or poor.

◆ ◆ ◆

It all began when biologists first reported a dramatic drop in tiger numbers in the late 1960s. They called for a total ban on tiger hunting and the export of skins. But banning tiger hunting was sacrilege in a land synonymous with the sport. It seemed impossible. Shikar was an entrenched industry supported by ex-maharajas, tour guides, hoteliers, travel agents, and a wealthy international clientele. They were vocal, aggressive, and well funded.

Despite strong protests, Gandhi pushed ahead. With a supportive legislature she enacted a total tiger hunting ban in 1970. But the months prior were some of the worst for the tiger. As rumors of an impending ban spread, people came from around the world to bag their last striped trophy. Rules and regulations were broken as hundreds of tigers, maybe more, were shot.

The shikar industry, still confident of victory, then went to the Indian Supreme Court to try to reverse the law in early 1971. The room was packed and tense on judgment day as the fate of the tiger hung in the balance. On the one side were the shikaris who claimed the ban would lead to job and foreign revenue losses, something a developing country could not afford. They claimed the tiger was not endangered at all and that biologists purposefully misled the government to enhance their own status and interests.

The government countered with witnesses who described the violent excesses of tiger hunting. Illegal poaching was rampant, and even legitimate hunters used their wealth and clout to surpass legal limits. Many witnesses were old tiger hunters who decried the complete collapse of hunting protocols. Hunting was no longer sport, but pure bloodlust. Scientists gave the strongest testimonies, arguing that tiger populations could no longer withstand such an assault. The human population was growing rapidly, and if tigers were to survive in India they needed protection from hunters and large areas of protected habitat. They persuasively argued that tiger conservation was a matter of national pride and prestige. India is the land of the tiger. The judge, moved by these passionate testimonies, ruled in favor of the cat.

A historical precedent had been set. The rich and powerful had lost to conservation, which had become a national political issue. Most importantly, it was recognized, albeit begrudgingly, and even by the shikaris, that the old relations between tigers and people in India had changed, forever. The tiger was no longer the ubiquitous forest creature, a living legend so common it could be shot at will. It was now endangered—a new concept that struck at the heart of a country so long associated with this animal and its wild forests.

The government launched a nationwide tiger survey in 1972 to confirm the scientists' findings. Only 1,863 tigers remained in India—depressing news, even for usually pessimistic scientists. The tiger crisis was now official. The World Wide Fund for Nature recognized India's struggles and contributed one million dollars toward tiger conservation, the funds that laid the foundation for Project Tiger.

But the hardest part had just begun and continues into the present. And that is setting aside habitat for the animal. Habitat loss to agriculture and industry has always been a bigger threat to tigers than hunting; stopping hunting was only a preventative measure that had to be enacted before the more arduous process of establishing reserves began. Creating viable reserves entails political and economic conflict. Rural villagers must be resettled away from their ancestral lands to give tigers and other wild creatures some peace and protection. Worse yet are the battles with powerful timber, agricultural, and mining interests who have little desire to sacrifice profits for the sake of a cat.

Gandhi's early conservation victory gave her the momentum allowing for the creation of a network of solid reserves. Project Tiger grew rapidly from 1973 until her death (by assassination) in 1984.[3] By this time the tiger population also peaked. These were the golden years for Project Tiger and the tiger. Unfortunately, they did not last. Ironically, since the death of Indira Gandhi, the tiger's greatest political champion, the tiger's fortunes have steadily declined.

Project Tiger was officially launched on April 1, 1973 at Corbett National Park in the Himalayan foothills. It was a fitting venue, as Corbett is the crown jewel of Indian national parks. Known for its spectacular scenery, it is particularly rich in wildlife and has one of the highest tiger populations in the country, estimated at over 100.

Corbett also has historical significance, as it was India's first national park. But it is most famous because it is named after Jim Corbett, the man who probably spent more time with tigers and knew more about the cats than any person who ever lived. Indira Gandhi was the political catalyst that made Project Tiger law, but the spirit that animated tiger conservation in India was Jim Corbett's.

◆ ◆ ◆

Jim Corbett grew up in the generation before Kenneth Anderson and was the first to write about tigers for an international audience (rather than just the English). He was born a few miles from Ramnagar in Nainital, located on a cool hilltop that served as a summer refuge for the British during the long Indian summers. It was an idyllic world for a young boy, and Jim spent his childhood learning about and exploring this vast natural playground.

He had identified almost all the birds of the surrounding forests (a considerable accomplishment considering there are over 400 species) by the time he was a teenager. He identified them by sight and sound and had a collection of dozens of stuffed specimens. Jim could also imitate wild animals. He was once practicing his leopard call and attracted both an English hunter and a real live leopard—to the surprise of all three.

Jim, like many boys of his generation, loved hunting. But it was born out of necessity. He had to provide food for his family after his father died when Jim was young. He used an old shotgun with only a single functioning barrel, a handicap that greatly improved his shooting and tracking skills.

Jim also learned about tigers early in life. As a child, he once heard rustling in the bushes behind his house, went to look, and found a tiger. The startled cat looked him in the eye and both stood in a trance for a few moments, before the tiger dashed off into the woods. Corbett recalled this encounter many years later, as it had made a permanent impression on him. The tiger became a lifelong obsession.

Corbett never married, and after a career with the Indian railroad that took him across the country, he returned to Nainital to retire, sharing the family home with his also-unmarried sister. He remained in India until it gained independence but, fearing reprisals against Europeans (which never materialized), left for East Africa, where he died at age 80 in 1955.

Like Anderson, Corbett was not a trophy or sport hunter; he shot man-eaters and other rogues who imperiled human life or killed livestock. He also had a unique hunting ethos, preferring to track on foot and only using a machan, which he considered unsporting, when absolutely necessary. It was rumored he often carried only a single bullet on his long jungle forays, which lasted weeks at a time, to even the odds further. And he never accepted rewards for his dangerous and painstaking work.

Unlike sport hunting for tigers, which was largely social and ceremonial, pursuing man-eaters was arduous and solitary work. These killer cats were keenly aware of human ways, making them not only dangerous but also extremely cautious and difficult to track. The hunter also had to be sure he was pursuing the right animal, because forests still teemed with tigers during Corbett's day. It sometimes took years before a man-eater was killed.

When a human victim was located, the hunter had to stand watch over the corpse until the killer returned. This macabre ritual lasted many days and nights. It became unbearable in the dry season, when temperatures hovered over a hundred degrees and the hunter had to hold his position in the face of barraging insects and the intense odor of putrefaction.

Emotionally, it was even more difficult. The mourning, often hysterical, family had to be restrained from removing and cremating the corpse, so that the cat would return and be killed. Since man-eaters remained at large for months, or longer, pressures on the hunter increased as death tolls mounted. Corbett once suffered from mental exhaustion after futilely pursuing the man-eating leopard of Rudraprayag for months. He lay out for almost a year, during which time the cat claimed 10 more lives.

Man-eating is unimaginable to the modern mind. But it occurred throughout India in the past. Only a tiny fraction of tigers ever became man-eaters, but when they did, they unleashed terror. A single man-eater put a halt to all activities—wood gathering, herding, farming, and travel—forcing people to hide in their villages. Entire villages were evacuated, sometimes for good, to escape the marauding cats. Man-eaters stalked the village outskirts for unwary passersby, while more aggressive animals simply seized their helpless victims right out of their homes. Some cats were so crafty they could enter a hut of sleeping people, seize their victim, and leave without waking anyone—another reason people thought tigers were phantoms.

Whenever a man-eater was shot it was a time of great relief and rejoicing. Word quickly spread across the countryside, and people came from miles away to behold the carcass of the fallen animal and to pay their respects to the tiger slayer who restored their lives normal.

Naturally, Corbett shared in their joy, often leading the procession, with the dead cat tied to strong branches carried by men into the village square. But he also felt regret, always believing the tiger was a "large-hearted gentleman," a noble creature naturally shy of humans. Man-eating was an aberration caused by disease, old age, and, especially, injury. Ironically, sport hunters, many of whom were amateurs, caused many of these injuries. They maimed rather than killed, forcing the cats to pursue easy prey like people or cattle.

Some injuries were natural. When Corbett skinned the Mohan man-eater he pulled 30 porcupine quills, some 5 inches long, from the animal's foot. The injury probably happened when the cat was young and not skilled enough to properly "flip" the porcupine before killing it. The flesh beneath the tiger's skin, from foot to chest, was oozing with soapy dark yellow fluid. While stalking the cat, Corbett noticed the animal moaned in pain with every step, a terrible handicap that forced it to become and remain a man-eater its entire life.

Corbett was a great hunter because he was also a great naturalist. He knew the forest and it inhabitants intimately, especially the tiger. He knew the cat's habits and how to read its mind. Successful pursuit required thinking like a tiger and anticipating its next moves, especially the place of attack. As such, the hunt became both a psychological and a physical duel between man and cat. Both stalked one another, moving into position among the light and shadow of the forest, behind rocks and trees, up hills, through valleys, across rivers and streams, each hoping to gain the advantage that could spell life or death—for either.

Why did the stalked cat not simply flee? Perhaps it thought it could make a meal of its pursuer. Or perhaps some competitive urge had been tapped, as the cat had met its match, a cunning human who could compete with the wily feline. Maybe there was some deeper primordial kinship between two hunters. Or perhaps it was simply the cat's fatal curiosity, which more times than not gave Jim Corbett the advantage.

Corbett's intuition also made him aware that the destruction of forests along with unchecked hunting would decimate tigers. He noticed them becoming less common, more secretive, less daring, and he was one of the first to warn that the tiger population would drop below 2,000 animals if habitat was not saved and reckless hunting not controlled—and that was in the 1940s. Despite his fame and knowledge, his words were not heeded. No one believed the tiger could ever be killed off. Corbett proved them all wrong. His warnings were remarkably prescient, because 20 years later the tiger population plunged—exactly as he predicted.

Fortunately, a few farsighted men in the colonial administration did listen. They recruited Corbett to organize and map out Hailey National Park, named after the provincial governor of the United Provinces, Sir William Malcolm Hailey, in 1936. (It was later renamed Corbett National Park.) Soon thereafter Corbett gave up tiger hunting and turned to photography to learn more about tiger behavior, producing one of the first films about the animal.

When Corbett left India he took with him the memories of tigers and India, which he then put into the books that made him world famous. He never returned India and his beloved tiger forests.

Corbett Park is in the terai—a rich blend of hills and valleys, forests and grasslands, abundant water and cool climate—a transitional zone between the freezing Himalayas and the scorching plains. The "terai arc," as biologists call it, is only a few hundred miles wide, but it stretches the entire length of the subcontinent, from the Pakistan border through Nepal and Bhutan, until it is absorbed by the Southeast Asian rainforests.

All of India's largest animals, including elephants, bears, leopards, rhinoceros, gaur, and water buffalo are concentrated in the terai. Naturally, it is premier tiger habitat. But terai tigers are noticeably different from other Indian tigers—larger, with thicker coats and bolder markings—more like the Amur tigers of Siberia.

These are the "royal Bengal tigers" of legend, named for their appearance and also because King Edward VII of England shot one in the eastern terai in the mid-nineteenth century, after which the name stuck. These big cats soon became the favorite trophies for sport hunters, making the terai the most desired hunting destination in India.

The terai is also known as the gateway to "Shangri-La"—the fabled mountain kingdoms of Kashmir, Punjab, Bhutan, Nepal, and Tibet, lands known for their scenic beauty and places of spiritual discovery. Today much of this ancient gateway has been lost to farmers and loggers. Yet, fortunately, a few oases like Corbett National Park remain.

Corbett is only 200 miles from New Delhi as the crow flies. But crossing the densely populated Gangetic Plain and climbing the winding mountain roads takes an entire day, at least by public bus, which I took from a crowded filthy public lot early one morning.

Most time was spent stopping to pick up passengers. Just like Indian trains, there is no exact point when the bus is deemed full. People kept packing in until there was no more room, and then they moved to the roof or hung out open doors. Finally, following the driver's instinct or simply the laws of physics, the bus took on no more.

We finally shed passengers in the less populated foothills. But the hilly, curvy road made for equally slow going. The only consolation was the scenery—mountainous throughout, with hill ranges lying in massive folds, growing larger, steeper, and more forested as they extended towards the snowy Himalayan peaks.

The lower hills were least attractive, with stunted trees and sparse vegetation thinned by huge herds of goats, cattle, and sheep, which ate their way though the undergrowth like locusts. Most belonged to nomadic Gujars who, like tigers and other wild animals, followed the seasons with their herds. During dry weather they camp near the rivers, then they spread out into the surrounding hills and valleys with the rains. The monsoon hits hard here, sometimes raining for weeks on end. But it creates explosive plant growth, much of it still present now in the late fall.

We drove ever higher, soon reaching the sal forests. The sal (*Shorea robusta*) is an imposing tree, with a heavy, dark trunk, spreading branches, and heights of over 120 feet. The massive sal groves evoke images of the mighty California redwoods. But these are deciduous, with deep green leaves, bigger than a man's hand, and showy yellow flowers that bloom in the spring. The sal is as useful as it is beautiful. The wood is used for lumber, incense, and oils; flowers and leaves provide fodder. But the sal is most famous because the Buddha was supposedly born under one—in the eastern terai, current-day Bhutan.

Sal dominates the forests of central and eastern India, just as teak does in the south. But the largest sal trees are found here in the terai, because of the cool, moist climate. They grow in stands so dense that they crowd out all other species. Some terai forests are over 90 percent sal. Yet their canopies, unlike the rainforests', allow sunlight to reach the ground—allowing shrubs and herbs to flourish, providing food and shelter to wild creatures.

The sal forests were now thick and lush, and I soon saw wild animals, mainly birds and monkeys. Then a few spotted deer dashed across the road, narrowly escaping the bus, which careened forward without pause. Only the ascending hills slowed our momentum, which was quickly regained on the descending straightaways.

Water became more abundant the higher we climbed. It trickled down the hillsides into roadside ditches. Streams flowed through every ravine and valley, winding their way among giant gray boulders shorn from mountainsides and mounds of cobbles worn smooth by the constant flow. These streams surge violently with icy gray water in spring, when the Himalayas thaw, only to retreat into small placid pools by the end of the dry season. They were now at an intermediate stage, flowing clearly and briskly.

Grasslands, or *chauds*, often covering hundreds of acres, grew in the rich soils left behind by the yearly floods. In some areas—especially the eastern terai—they become grass forests, spreading miles and towering 20 feet or more, making them the tallest grasslands on earth. They were about half that height in these parts.

Farming villages built of mud and thatch were scattered throughout the valleys, but they were always established above the flood zone. Some, made of stone, sat high in the mountains, surrounded by terraces that controlled erosion and captured water for wheat, root crops, and vegetables. Livestock were equally well adapted, cutting narrow trails into the steepest hillsides, where they stood, precariously, grazing on short hardy grasses.

Taller mountains stood in the distance, many completely denuded by constant grazing and farming. But other areas were naturally barren, with gaping holes and massive rubble piles the result of landslides. The Himalayas are young mountains still growing a fraction of an inch per year; this growth forces them to buckle and constantly erode.

The old bus strained and lurched the last few miles of the trip. But the swaying and constant slowing and accelerating took its toll. People started getting sick. First a little boy vomited on a suitcase a few rows up. Then another. Soon six or eight heads hung out the narrow steel windows to relieve themselves. The ticket master walked down the aisle, I assumed to help. To the contrary. He began yelling at people to shut their windows as vomit was flying back into the bus. And it was. Fortunately I had an aisle seat and sat next to a good interpreter who explained the drama as it unfolded.

Closing the steel shutters seemed to make things worse. It was dark and stuffy. I could not see anything outside; there were only sounds—the muffled belches and faint groans of discomfort. I peered ahead through the dusty front windshield and saw flat ground approaching. We had reached a summit. The old bus slowed and sputtered with relief, as did the traumatized passengers.

We were in the Ramganga River valley, home to the small town of Ramnagar, also park headquarters. Despite Corbett Park's fame, Ramnagar was not as touristy as expected. It is an old town, predating the park, thus it has grown organically and incrementally like any normal town. Corbett Park also has most of its lodgings inside its boundaries, which because of its larger size—some 2,400 square miles—can accommodate people and tigers.

Still, many hotels were clustered along narrow streets, and I scouted about for a suitable one for the night and soon found one. It had a hand-painted picture of a tiger in the front window with a small faded photograph, cut out of a book or magazine, next to it. It was an image of a man, lean and slightly balding, with a cropped moustache and sharp, expressive eyes. He wore khaki shorts and had a gun in hand and was standing next to a slain tiger. The famous Jim Corbett.

◆ ◆ ◆

I got a slow start that first morning at Ramnagar, as the cold Indian morning kept me in bed later than usual. I had actually been awakened before sunrise by a family of roosting pigeons who entered through a small hole in the corner of the high ceiling. After a few minutes of noisy cooing and prating directly above my head, I feared the worst and sat up to shoo them out with a meager yell. They simply flew to the other side of the room and continued their noisy display without pause until the light of dawn finally lured them outside.

I had barely gotten back to sleep when the blaring of car horns started in the street below. Ramnagar was awakening. I eventually got back to sleep, until a chai wallah (tea merchant) set up shop directly under my window and began yelling loudly to attract his equally noisy customers. Finally I gave up and went down to join the melee, eating a hasty chapatti breakfast from a nearby vendor followed by many cups of the wallah's syrupy tea.

Then I was off to park headquarters to get entry permissions. This took another few hours. The lines were long and there was only one attendant, who had the habit of getting into lengthy discussions with each customer. Many were local villagers who came to wrangle over park rules and fees, and the rest, tourists who had to gain official entry permissions.

After I got my papers in order I went to find a driver and quickly succeeded. His name was Karaan, a slight, middle-aged man, lightly complected and partly balding. He was polite, but somber, and spoke little. But his English was good, and he told me that he was also a good driver and cook and has known this forest for "many, many years."

But, as soon as I met him, he took off again. "Only a few minutes," he assured me. "I have to take my cousin to his mother's house. It's very close." It took an hour. He probably paid his respects to the family, undoubtedly a large one, and surely had a cup of tea or two. I did not fight it and just waited in the hotel foyer, drinking more tea myself until I was thoroughly wired on sugar and caffeine. I was ready for those serene woods.

He arrived and we packed up, but we still had to navigate the crowded streets, taking even more time. We entered the park just as the sun began to arc towards the horizon. It was perfectly quiet, as we were the only car on the road; the air was cool and breezy, but the sun still shone warm wherever it broke through the shady woods.

We drove miles through forested hills, stopping often upon my request, breaking park rules by wandering off side trails to get better views of the forest. The scenery everywhere was spectacular—steep mountains carpeted in solid green forest, separated by wide, cobble-strewn valleys flowing with clear water. At one point we stopped over a massive pool where numerous mahseer—some over three feet—drifted lazily under the water's surface.

A few crocodiles lay about on the shore. They were Indian muggers (*Crocodylus palustris*). Their rounded muzzle and stout, dark bodies, offset by lighter yellowish bands, made them similar in appearance to the American alligator. And like their American counterparts, they rarely grow to more than 12 feet but can still be dangerous. I turned and asked Karaan about muggers and people.

"Mostly they are harmless, but people are careless and bathe or wash at a stream and are attacked. A teenage boy was killed recently a few miles away. He was out playing in the water and the crocodile pulled him under. His sister saw it and went screaming for help. But it was too late," he explained. "They were careless," he added nonchalantly.

"What happened then?" I asked.

"The villagers found only a few body parts. Naturally they wanted revenge." He turned and told me in an authoritative voice, "You can't kill them. They are all protected."

The few down at the pool seemed pretty harmless, as all were less than eight feet long and rather lethargic, despite the late afternoon sun. Karaan told me a government restocking program that has helped bring back the species at Corbett and other sanctuaries around the country released them here. Today India is one of few Asian countries where the mugger is not endangered. But as their numbers rise, so do conflicts with people.

The road followed the stream a while longer then veered off in the opposite direction. We pulled off at the juncture, in a place overlooking the water as it flowed off into a wooded ravine. My driver jumped out and walked to the lookout point while I fumbled in my bag for my binoculars. "Come quick," he shouted under his breath, hands waving frantically. "A tiger, a tiger! Come quick!" I dashed forward as he pointed to the edge of the forest near the stream. "Look there! A mother and cubs."

They were far off and I missed them at first. Then I saw the last cub, just as it jumped up a hill following the other two as they disappeared into the woods. But there it was—a tiger!

We were both elated and stood watching for some time, thinking they might come back out of the forest.

"I have seen many tigers since I have worked here. But I am always so happy when I see one again," Karaan exclaimed. "I am so happy every time," he proclaimed again. He even mustered a slight smile, and then we both stared hard towards the stream hoping they would return. We sat for a half hour, hoping, waiting. But the cats were long gone.

The sun now lay low on the horizon, so we headed back to our jeep to get to our destination before dark. We drove a more few miles until we reached an old forest bungalow that was off a small gravel road and right next to a small brook, filled with boulders.

Karaan prepared our room and started dinner as I went out to sit on the front porch, gazing at the water and the dark forest. Night fell quickly, so I listened for forest sounds. It was too cold for most insects. The only sounds were the soft, gurgling waters and the occasional unidentifiable screech or rustle from distant trees. A solitary nightjar and a few bats fluttered close by. The moon, almost full, shone brightly, occasionally blocked by high, wafting clouds.

We ate dinner late and I slept well, although in a cold room and on a hard wooden bunk. But there were no pigeons, chai wallahs, or blaring car horns. Only the quiet of the sal forest.

◆ ◆ ◆

I spent the next several days taking forest excursions with Karaan from our bungalow base and saw many parts of the park and familiar animals—deer, monkeys, birds, crocodiles, even a small herd of elephants—but no more tigers. Karaan then suggested we go to Dhikala, the main tourist complex, to take an elephant ride into the forest. So we drove out to the compound to arrange for an excursion the next morning. It housed a few dozen people and had shops and restaurants—a comfortable bourgeois oasis in the middle of the forest.

We arrived at Dhikala the next day before dawn. It was quiet except for a handful of people, mainly Indians with a few Europeans and Americans, who stood waiting for their elephant ride. The animals soon marched up to a nearby platform, which we climbed to get on to a thick pad that was secured on the elephant's back. A mahout sat on the animal's neck and guided it with his feet. He also had a heavy steel ankusha—a rod with hooks that grabbed the animal's ears, to make it turn, and a sharp point, to prod it go faster.

Mahouts are not just elephant drivers. They are caretakers who feed and house the animals. Bathing is one of the most important responsibilities, especially during dry, hot weather when elephants must bathe daily to cool down and protect their skin from drying. Being a mahout is a family tradition, often caste-bound, but there are also many Muslim mahouts, especially here in northern India. Children of mahouts begin by caring for baby elephants, often eating and sleeping together, thus creating bonds that last a lifetime.

Our mahout was an old man, with a long white beard and deep-set eyes. Like the other mahouts he was small, but wiry, like a horse jockey. He knew no English, or if he did, he never spoke it. In fact, he never spoke to anyone but his elephants. He was stern and steadfast and ready to move. A German and a woman from California joined us. The German took pictures while the woman turned and talked to me as we separated ourselves from the other elephants and moved towards the distant forest, still shrouded in early morning mist.

"I love India," the woman exclaimed as we entered the woods. "It all started with my first yoga class. Yoga taught me how to free myself from myself, to become one with the world. People here are so much freer than Americans. They understand peace and tranquility and beauty." She moved her hands gracefully as she spoke, her head tilted upward as she gazed, trance-like, into the sky. "We Americans are so materialist and violent. We could learn so much from Indians."

At that moment the mahout began furiously hitting the elephant with his iron rod, to force it through the dense underbrush that we had just entered. The woman let out a slight shriek. "He is hurting that animal," she intoned with indignation. The mahout kept whaling away, hitting the animal with all his might. "Why does he keep having to hit that poor creature?" The mahout continued.

"I think elephants are used to this." I tried to explain. "After all they have really thick skulls." She remained unconvinced, even crestfallen, the image of Indian peace and tranquility shattered by this little man with an iron rod. I tried to console her. "Look, he is a Muslim, not a Hindu. I think their idea of peace and tranquility is different." This seemed to assuage her a bit. The hitting finally stopped as we made it through the foliage. She gave a sigh of relief but looked at me pensively, shaking her head.

The German, meanwhile, had put away his camera and was busy collecting leaves from the 10-foot cannabis plants that now surrounded us. He told us he had just come from Singapore, where they will put you in jail for possession. He grabbed another huge handful, stuffing it into his coat pocket.

"Yes and they probably will cane you as well," I added.

"Yeah, like an elephant," he laughed loudly. The woman was not amused.

Many Westerners think Indians commonly smoke ganja while practicing yoga and Kamasutra. Ganja, at least, is usually associated with forest tribes and lower castes. A self-respecting Brahmin would never smoke ganja. This German was no Brahmin. He stuffed both coat pockets full and would have continued, but we left the pot patch for a forest clearing, where we stopped and waited.

Karaan whispered that deer come here to graze. Naturally tigers follow them. We waited a good half hour for activity, but nothing stirred this cold morning. The German pointed out an eagle in a distant tree. It was a hawk eagle (*Spizaetus nipalensis*), a large brown and white bird. It sat motionless, head tucked deep into its warm fluffed-up feathers, with its characteristic black feathery crest barely visible. We watched it for a while, hoping we could see it in full flight. But it remained put. We moved on to the Ramganga River, a few hundred yards off, to see what we could find there.

We stopped to search for life along the riverbank. Nothing stirred, so we crossed. The elephant stopped to drink as we entered the water. The mahout prodded it to continue, then yelled commands and hit the animal again. The Californian tried not to notice. Karaan, sensing her unease, explained that not all mahouts are this harsh. "They are all different. Maybe this is a difficult elephant. They too have different personalities. Sometimes they are very stubborn." She resigned herself to the situation. Perhaps it was simply the elephant's fate, or perhaps hers.

We crossed and entered a large sea of grass, wilted this time of year but still thick and tall in many places. We moved slowly, so as not to scare any animals. Soon there was rustling and movement a few yards ahead. We saw only the quivering of the tops of the grass as some unknown creature bolted away from us. "Maybe a hog deer," Karaan said. "They like this heavy grass."

The hog deer (*Axis porcinus*) is about the size of a spotted deer, but not as attractive. It is, as its name suggests, a short squat creature, with a nondescript brown coat. It is rarely ever seen because it lives a secretive life in the foliage along waterways. The animal is also becoming rare throughout India as this type of habitat dwindles.

We reached shorter grass where a dozen spotted deer grazed. They did not even flinch as we approached. Most animals, even tigers, allow elephants to approach them, not noticing the human passenger on board—an advantage of viewing wildlife on elephant back (and one hunters used in the past). But spotted deer do not fear people in general, as many live in and around farms and pastures. We moved closer.

The German whipped out his camera and began snapping away. The rest of us, except the mahout, admired the total experience. The grazing deer moved slowly, gently nibbling at the yellow grass. Behind them were taller grasses that formed a curtain through which the river flowed. The pale dawn sky was now a tantalizing blue, turning the purple-black mountains deep green and melting the mists that hung low in the valley.

The Californian looked at me and beamed, her faith somehow restored. The gateway to Shangri-La—or some other beautiful place—lay wide open.

◆ ◆ ◆

I never saw the woman from California again but did encounter the German at Dhikala a few days after our elephant ride. I had gone back there to pick up a few things and relax before my arduous return to New Delhi the next day. I sat at an outdoor tea stand, taking in the warm morning sun and reading a book by Jim Corbett—*Jungle Lore*, one of his last—that I had just bought at the gift shop. It was a counterfeit copy, like many foreign books printed in India, poor in quality, pages falling out, but cheap and something to read.

I almost went out to the large watchtower that overlooks the Ramganga a few hundred yards behind the tourist complex, rather than sit and drink tea. I decided against it after seeing troupes of tourists moving in that direction. It would be crowded and noisy. How could that be enjoyable? How could that attract wild animals? I would see no tigers. What a waste of time—so I stayed put.

I had been reading and drinking tea and watching jackals, completely unafraid, move across the open lawn in front of me, raiding trash cans and searching for food scraps. Wily as coyotes and deft as foxes, they are good hunters, but scavenging after tourists was an easier, more secure livelihood. I saw no fewer than six moving about.

Then along came the German. He recognized me from afar and waved, walking briskly towards me. He continued waving and grinned broadly, obviously eager to tell me something.

"I just saw a tiger!" he proclaimed breathlessly.

"Where?" I asked in a mix of amazement and envy.

"At that damned tower," he said in disbelief. "It was a huge male. He came right down to the river to drink and just sat there for a long time."

"You mean he just sat there in front of you and all those people? He wasn't scared off," I asked.

"No. Here, look at my pictures." He pulled out his camera. He shook his head, still in disbelief as we looked at the images. "I spent all that time and money going on all those damned elephant rides and I see a tiger right here out on the platform surrounded by screaming children and a bunch of noisy tourists. I can't believe it."

Neither could I. The mysterious tiger.

We talked for a while longer as he showed me the rest of his pictures. It was almost noon. Kristof, as he was named, then bid me goodbye as he had to ready himself for his trip back to New Delhi.

I was almost tempted to head out to the platform myself to use this last afternoon to see another tiger. A wise and efficient use of time, I thought. I will maximize all my opportunities to see a tiger while I am here. Then I balked as I saw a large train of Indian tourists moving in that direction. It was lunchtime, and "picnicking" is a favorite Indian pastime when visiting the countryside. It is especially popular at national parks, where many people simply come for the day to eat lunch with their extended families.

I was again reminded how different their idea of nature (and simply life) was from that of most Westerners. Even as tourists, we come as explorers, even conquerors, covering vast areas searching for tigers or ruins or cultural relics of some kind. Or we visit museums, all of them, and sample as many varieties of food as possible and buy trunk-loads of artifacts to validate our experiences. "Been there, done that." The great race to see as much as possible before we die, or before it all ends.

The Indians meandered into the distance, noisy, blissful, carrying their big baskets of food, with aunts and grandmas and children trailing behind. Maybe they would see a tiger. Maybe not. Few would care if they did not. After all, it was fate.

I sat back down to read Corbett and was soon entranced and transported. His images were now far more vivid since I had experienced them firsthand. I glanced at the surrounding mountains and the river off in the distance, nodding in agreement as his vision merged with mine.

Corbett's descriptions of forests and wild animals were as much a hallmark of his writings as was hunting. He also wrote in an autobiographical and elegiac vein, especially in his later works, praising what he called "India's long suffering people." They endured poverty, disease, ruthless rulers, and the predations of wild animals. Yet they had an indomitable spirit and a humility that he found refreshing, especially from the routines of officialdom which he knew all too well.

This made me think of Project Tiger, that immense government body with the "official" mandate to protect tigers. Did it, like all bureaucracies, succumb to its own contradictions? This bureaucracy located in the middle of one of the world's largest cities was as far removed from tigers and nature as possible.

Maybe the secret of tiger conservation was not just with bureaucracies and the specialized knowledge of armies of "experts." Maybe it lay elsewhere. Perhaps even with the Indian people and their culture.

— 5 —

Tigers and Gods

Some conservationists argue that protecting tigers requires more than science and government; it requires culture. If conservation principles can be integrated with cultural beliefs and practices, they will be more intelligible to the common person.

A cultural approach is particularly well suited to a country like India, with its ancient traditions and widespread religious devotion. Despite the divisions and conflicts between its many faiths, castes, and ethnic groups, all share a general respect for nature that allowed teeming populations to coexist with wild animals in more places and for longer periods of time than almost any other place on earth.

I had already seen some sanctuaries using religion to convey conservation principles to villagers. At Bandipur I remembered seeing a forester telling villagers that forest destruction was offensive to the local forest god. It was more effective than trying to explain biodiversity loss and sustainability. The language of science had little impact, so the ancient language of the gods had to be invoked.

I was reminded how deeply nature and culture were interwoven when I returned to New Delhi, even in this sprawling, crowded city of 15 million people. Streets and marketplaces exuded the aromas of incense, curries, and burning wood. Women's saris reflected the vitality of the forest and their jewels the mysteries hidden in the earth. And animals were an intimate part of life everywhere.

The sacred Hindu cow was ever present. Confident of their unassailable spiritual status, cows slept and strolled and ate and defecated wherever they pleased. They were joined by goats and sheep that grazed in grassy patches alongside busy roads and open lots near high-rise apartments. Snake charmers plied their trade in parks for tourists or brought their poisonous commodities to town for religious festivals and rituals.

Even wild animals thrived in the heart of the city. Mynahs and flocks of lime-green parrots gathered high in the treetops, and tiny sunbirds moved about the tangles of vines and shrubs that aggressively colonized any patch of open land. Honey buzzards soared high in the sky on four-foot wings, taking advantage of the streams of warm air emitted by the concrete and asphalt below. Colorful agama lizards raced up and down walls and tree trunks, avoiding hungry birds or chasing insects that buzzed and crawled and wriggled about. And in the evenings, small gray geckoes gathered around lighted areas to stalk moths. Crickets and a host of unidentified insects sang uninterrupted until dawn.

Wild tigers live nowhere near New Delhi, but they still run free in the Indian imagination. As I walked and rode around the city I saw many pictures of the big cat. Tiger books were prominently displayed in bookstores, and tiger pictures appeared in restaurants and were plastered across vendors' stands and painted on the backs of taxis and trucks. Growling tiger faces and sleek tiger bodies were used by merchants to sell everything from shampoo to insurance.

The tiger's strongest presence was in religious imagery that appeared in everything from postcards to giant murals on walls and buildings. Sometimes tigers were joined by other animal images like Ganesh, the god of knowledge, with his elephant head, or Hanuman, the monkey god of hope and devotion. When displayed together, they created a panorama that made the forest come alive on street corners and in other public places. The ancient forests were gone here in the heart of the city, but the tiger and other wild creatures still have the power to invoke fear or reverence. Could this spirit be harnessed to protect tigers?

I left New Delhi and traveled eastward through the entirety of the Gangetic Plain to Kolkata (Calcutta) to gain greater perspective on this question. After all, this is the heartland of Indian civilization—home to many of India's religions and ethnic groups, where nature and culture have been intertwined for millennia, and one of the first places tigers entered human culture and consciousness.

I got to the train station at mid-morning so it was not terribly crowded, allowing me to buy a ticket and get out to the platform in good time. But when the train rolled in an hour later, the platform was filled to capacity. There was a short mad rush once the doors opened, facilitated by small armies of coolies wearing weathered red vests, which designated their official status as baggage carriers. They ran about frantically looking for anyone with bags and then whisking them aboard as quickly as possible, with the hopes of securing another client before the train left. After 30 minutes of bedlam, the train began its journey.

Indian trains are rudimentary machines whose designs have changed little for decades. They are solid steel throughout, with no style or ornamentation. Their purpose is simply to withstand hundreds of millions of passengers every year without falling apart. Even inside, there are no frills or comforts, not even a dining room. Meals are cooked in a special kitchen and then served to passengers in their seats. Most people bring their own food. Even my second-class car was basic.

The only advantage it had over third class was a guaranteed seat, large glass windows (as opposed to steel bars), and air conditioning, which was largely ineffective because most windows and doors were left open. It was also closed off to vagrants and beggars who jump aboard third-class cars at every stop to ask for food or money.

The first few hours of the journey were slow because the train halted often in the many towns and villages that radiate outward from New Delhi. The clamor and confusion was repeated at every stop as people struggled to get aboard. Vendors and coolies moved deftly around the flotilla of elbows and arms, gesturing wildly and yelling in shrill tones to sell or carry what they could before the train moved on. The struggles were most acute to get into the third-class cars, as people pushed in through the narrow doors to claim any seat. But seats quickly became unnecessary as the cars filled to standing room only. When all space was taken on the inside, passengers simply climbed on the roof, trusting fate that the train would not veer wildly or brake suddenly en route.

We eventually entered the open plains, a patchwork of fields, pastures, villages, and temples tied together by ancient footpaths and now modern roads and rails. Poverty is as common here as in the city, but it is buffered here by bucolic surroundings and a pace of life tied to the movements of the sun. Large herds of cows still wandered unrestrained by fences or, it seems, unclaimed by any owners. They were joined by goats, sheep, and the formidable black water buffaloes that do much of the work in rural India. White egrets and brown bitterns congregated in wet fields, and large blue Indian rollers sat high in the palm trees looking for unsuspecting insects below.

The plains were perfectly flat and stretched as far as I could see, created by silt deposited by the mighty Ganges and its countless tributaries for eons as they wound their way eastward before emptying into the Bay of Bengal. The Gangetic Plain covers thousands of square miles, and its rich alluvial soil reaches depths of over 20,000 feet, making it one of the most fertile places on earth.[1]

This tremendous river system is created by the monsoon, which sweeps up off of the Indian Ocean every summer and drenches the entire continent for months. The rains then ascend the Himalayas, where they fall as snow. These snows, along with massive high-altitude glaciers, then thaw during the dry season to provide the continent with water until the rains arrive again.

Ample water and fertile soil is what makes the Gangetic Plain so densely populated. The Plain, together with the neighboring alluvial plains of the Indus River valley in Pakistan and the Brahmaputra River valley in Bangladesh, is home to one sixth of the world's people. The soil here is so fertile it is estimated that it could grow enough food for several hundred million more people!

It is hard to believe that this land was once heavily forested. Today almost all are gone. The only trees I saw were clustered around villages, mainly fruit trees. Small groves of wild species lined stream banks, and a few solitary trees stood in open fields, creating oases of shade for cattle and field hands.

As the train raced across the open landscape I tried to imagine what this ancient forest world must have been like. It was a composite of woods and grassland, interspersed by streams and watery depressions, making it perfect wildlife habitat. Historically, the number of large mammals found there was rivaled only by that of the great Africa plains.

Imposing herds of elephants moved slowly through mist-filled forests, plundering and feeding upon the succulent undergrowth that was also home to bison, bears, and leopards. Water holes marked low-lying areas and were hidden from view by towering grasses. They were secretive haunts of rhinos and shiny black water buffaloes, whose 10-foot racks made even tigers wary.

Swamp deer (*Rucervus duvaucelii*) congregated in open areas by the hundreds, moving in perfect and graceful unison until a tiger came into view, causing them to split into dozens of wildly weaving lines that merged after a great distance when the threat had passed. Abundant water, prey, and cover made this the perfect tiger haunt. Countless thousands prowled this land, uninhibited, for millennia.

This flat land also formed a bridge between Asia and Africa. According to plate tectonic theory, it was created when India, then an island, crashed into the Eurasian landmass to create the Indian subcontinent about 10 million years ago. India's location between Asia and Africa made it a center of animal and plant speciation, which is why it has one of the highest levels of biodiversity on earth today.

The lion, cheetah, and antelope came into India from Africa and the Middle East and inhabited the drier zones of the subcontinent. The rhino and elephant, deer, bear, and the tiger entered from Asia and inhabited the wetter zones. As such, India is the only country on earth that is home to "lions and tigers and bears."

The land bridge also brought people. Indian history is a history of invasion and migration, with each successive group of people taking advantage of abundant resources and each incorporating nature into their respective practices and beliefs. The first human inhabitants in India were tribal peoples—hunters and gatherers—who entered slowly and piecemeal, from both east and west. This is why tribal people are also called *adivasis*, which means "original inhabitants" in Hindi.

Tigers had a strong presence in tribal culture, because these people had the longest and most intimate contact with the cat. This is the case with all forest tribes throughout the tiger's former range from China to the Caspian Sea. But in India, tiger myth, ritual, and worship was particularly intense and widespread because tigers and people lived together in virtually every habitat across the land.

The Naga of far eastern India believed man and tiger were brothers, both created at the beginning of time and born of the same mother. Naturally tigers took on human characteristics, and people took on those of the tiger in Naga legends. It is rumored there are still "were-tigers," who can change from tigers to men, and "tiger men," who have scars on their bodies purportedly corresponding to the wounds of real tigers living in the forest.

The Warli of western India believe the tiger god Vaghadeva is the greatest of all spirits, giving life to every living thing. Tiger statues guard the Warlis' fields, ensuring fertility to women and the soil and protection from evil. Vaghadeva is closely related to the goddess of marriage, and the tiger is the first spirit to be invoked during a Warli wedding ceremony, in which the couple wears red and yellow to symbolize the tiger.

The Gond tribe of central India has a special altar dedicated to the tiger goddess Waghai Devi, in a spot where a Gond woman was once snatched away by a tiger and taken into the forest, never to be seen again. The Bhils of neighboring Rajasthan believe all men are descendants of tigers. They worship the tiger prince Waghaika Kunwar and frequently offer him fruits, liquor, and livestock.

Tribal communities have remained small and their technology simple, making their overall impact on the environment and the tiger negligible. If any culture has lived in harmony with nature throughout history, it is these ancient tribes. The next group of invaders—the Aryans—was different. They were warriors, farmers, and pastoralists, bent on conquest and control.

The Aryans first entered India almost 4,000 years ago from their mountain homes in central Asia. Over the next several centuries they spread agriculture, the caste system, and the Hindu religion—first across the plains and then into the heart of the country.[2]

Hinduism is an ancient and complex religion. It is the only one of the world's major religions that can be traced back some 5,000 years to the very beginnings of agriculture, or the Neolithic Revolution, as it is often called. Many scholars think that Hinduism is a composite of ancient Aryan beliefs and existing tribal religions. As Aryan power spread, it became more unified and coherent.

Hinduism is a mystery to most Westerners. Many find it alien and blasphemous because of the belief in reincarnation, its many gods, and the exaltation of suffering. But to others, Hinduism is exotic and alluring. It is associated with meditation and pacifism, with vegetarianism and ornate temples, or with yoga and gurus who can return stressed and materialistic Westerners to lives of inner peace and harmony. Whatever the outsider's view, Hinduism is a truly living faith that influences every facet of life in India.

But Hinduism did not begin with peace, harmony, and beauty. When the Aryans first entered the Gangetic Plain they exploited tribal peoples and their forest homes, dramatically altering the natural landscape. They had huge ritual forest burnings, or *yagnas*, which destroyed hundreds of acres at a time. Priests doused the woods with vats of animal fat or ghee from sacred cows, and as the flames roared away, they invoked the blessings of Agni, the fire god; Indra, the sky god; and Vayu, the wind god; who together symbolized the great forces of nature, which had to be harnessed for their civilization to thrive. Even animal and human sacrifice commonly accompanied these massive forest burnings. The forest gave life and it took life—it was the will of the gods.

One of the greatest forest burnings is depicted in the Mahabharata Epic, where the Khandava forest is burned to the ground during the war between two rival tribes. In the story it burns so intensely that all animals are killed and even the gods think it signals the end of the earth. Khandava is now gone, covered by present-day New Delhi, not far from the train station from which I departed.

However, this destruction reached its limits about 2,000 years ago, as mountains and malarial swamps formed natural barriers to human settlement and cultivation. The Hindu religion also changed. The rituals and beliefs that celebrated nature's destruction were replaced by practices of protection and veneration. This was best captured in the Hindu idea of balance, or dharma, where the desire to control one's life or natural surroundings is minimized—an attitude that undoubtedly benefited the tiger and his habitat.

Tigers and other wild creatures also became objects of veneration and were associated with spiritual and moral principles. The goddess Durga, the great mother goddess who rides a tiger, protects people from evil and preserves moral order. She has eight arms to carry the weapons that help in her eternal struggle, and the tigers are naturally her vehicle, because they represent the unlimited power of goodness.

Shiva, the destroyer, is always seated on a tiger skin, which in this case represents the potential energy of creation. Since creation always follows destruction, the tiger is associated with both forces. The tiger is also equated with pride, which must be overcome ultimately in order to unleash the goodness of creation.

The train pulled into the train station in the city of Agra that afternoon, where I was soon greeted by clanging of metal and a loud, mantra-like "Chai, chai, chai." It was the familiar chai wallah, carrying a large dented teakettle with a ladle hung on the side to serve customers. He held it with an old cloth rag wrapped around the handle to protect him from the heat. He moved briskly down the aisle, and I caught him just in time to get a tiny plastic cupful, so thin it barely contained the heat.

I walked out onto the crowded platform to see the surroundings. There was a large factory in the background spewing out thick clouds of smoke, and cars, buses, and rickshaws crowded the narrow streets, belching out their noxious fumes. Electrical wires and poles spread out in every direction. I looked up into the smog-filled sky and saw a few honey buzzards circling overhead, barely visible.

Agra is famous for the Taj Mahal, one of the seven "wonders of the world." It was also one of the most important cities in the Islamic civilization, which began to grow in India after the ninth century. Like the Aryans, the Muslims entered India from the west and gradually expanded their influence. Islamic power reached its height under the Moguls, who ruled for 300 years from the sixteenth to nineteenth centuries.

From their capital of Delhi they used their military might to create a swath of Islamic culture that extended from the Indus River valley in the west, across the entire Gangetic Plain to the east, enveloping the Brahmaputra River valley.

They subjugated or converted the Hindu population and continued the process of forest clearing and exploitation.

When the Taj Mahal was completed in 1648, Agra, Delhi, and other Islamic cities of the Mogul Empire were still surrounded by forests and wild animals. The famous Mogul hunters and their massive expeditions did not have to travel far to shoot a tiger. Even a short excursion to the outskirts of Agra could yield a full bag of fowl and antelope.

Mogul hunters in this, the drier part of the Plain, were renowned for falconry and coursing with cheetahs. Both creatures were raised and trained specifically to hunt. The Mogul emperor Akbar reportedly had 1,000 cheetahs in his palace, which probably had better care and conditions than most of his human subjects.

Mogul emperors were also enthusiastic naturalists who collected and studied wild plants and animals. They were equally enthusiastic art patrons and commissioned thousands of nature paintings during the height of the empire. Today these paintings are still considered some of the most scientifically accurate ever rendered.

The lion was always the symbol of Mogul rule, as it was the more common cat in the western parts of the country where their power was centered. But Muslim rulers who lived in tiger territory often adopted the tiger as their icon. The most famous was Tipu Sultan, ruler of Mysore, who saw the tiger as the perfect symbol for Islam and became a devoted tiger fanatic.

In southern India young Muslim men paint their near-naked bodies black and yellow during the annual tiger dance (Kalipuli), gyrating through the streets like cats to honor the Prophet Mohammed's grandson—who was seen to have the courage of a tiger. In the Sunderbans mangroves that straddle the Bangladesh-India border, Muslims and Hindus alike recognize and pray to a host of forest gods, including Daksin Ray, who controls crocodile and tiger spirits and is able to enter a tiger's body at will. He is always depicted holding some type of weapon and often rides on the back of a tiger, which is his vehicle.

The Moguls eventually capitulated to the British in the nineteenth century. But Islamic influence is still strong throughout the Gangetic Plain today, as this is where most of India's 150 million Muslims live.

Most are Sunnis. But they differ from those in other parts of the world. Muslim women in India wear colorful saris and often do not cover their heads, and Muslim cabdrivers burn incense and glue small statues of Hindu gods on their dashboards just like their Hindu counterparts. Many here are influenced by Sufism, a more mystical rather than legalistic branch of Islam. And like the Hindus, many Muslims remain under the tiger's spell.[3]

As the train whistle blew, I finished off the last bit of my milky tea and looked for a place to dispose of my cup. The only trash can available was the railroad tracks, so I put it in my pocket to take back into the train. As I walked along the platform back to my car I could see hundreds of little white cups lying under the train. Plastic is cheap and convenient, but not the best thing for India's rudimentary trash system, which can only accommodate things that will rot, wither, or be eaten.

The train whistle blew again as we rolled out of the station and moved slowly through the crowded city center. We passed an ancient mosque built in fine smooth stone, its pointy minarets reaching to the heavens, now clad in smog. We soon hit the outskirts of town—places where cheetahs once ran down antelope—now covered by congested slums. Feral dogs, nervously on the lookout for food or foe, patrolled the narrow dirt streets, and pigs wallowed in the open sewers that lined them.

We finally picked up speed as we entered the familiar patchwork of farms and villages. We soon passed a large pond where a boy had taken his herd of cattle to drink. I watched him yell and gesture to keep his herd together. A pair of cranes stood only a few yards away, completely unperturbed by the commotion. They moved slowly and gracefully in this place as they have for thousands of years.

◆ ◆ ◆

Two days' journey brought me to the city of Varanasi, located right on the Ganges River. Also known as Benares, it is Hinduism's holiest city and the birthplace of Buddhism. It is also one of the world's oldest cities. When Mark Twain visited India he wrote, "Benares is older than history, older than tradition, older even than legend, and looks twice as old as all of them put together."[4]

At first glance Varanasi is like other Indian cities. But its antiquity is revealed once inside the city center. It is built directly on the water's edge and on a human (and animal) scale, with ornate stone buildings clustered tightly together and connected by a labyrinth of narrow streets and alleys, also paved with stone. They are too narrow for cars, even rickshaws, but perfect for living things.

Cows were common, ambling about, blocking thoroughfare, and dropping manure everywhere. I was trying to find a hotel after I arrived in town when I rounded a blind corner and was almost flattened by a small stampeding herd. Fortunately for me, they made a sharp right turn and raced into the open door of a man's living room—all 10 cows. The owner came to the door and as he shut it waved and smiled, obviously proud of his animal wealth.

I wandered down to the water after finding a hotel to watch the hundreds who had gathered to bathe, swim, and meditate at the shore. They gathered all along the extensive steps, or ghats, that extend along the entire shoreline and lead up to the dense cluster of temples, hotels, and opulent aristocratic palaces built just above the high-water mark.

Most visitors here are religious pilgrims, who come from across India to pray at temples and especially to partake of the sacred waters of the Ganges. They bathe in it and collect it in jugs to take home. The sick and elderly come here to die so their ashes can become one with Mother Ganges.

Acrid smoke filled the air as I wandered towards the burning ghats—the place of cremation. Hundreds of human corpses are burned here every day and their ashes spread into the Ganges. I watched young men bringing constant supplies of wood to fuel the fires and human corpses to be consumed by them. The covered bodies were laid out in formation, all waiting their turn for their final earthly ritual.

Dogs and crows watched furtively from the sidelines, on the lookout for charred remains or anything edible left behind by the throngs of worshippers. The water near the shore was opaque, a disagreeable greenish black, because of ashes and the constant runoff from the crowded city above. But it was offset by a patina of color—hundreds of flowers, petals, leaves, and other detritus floating on the surface—left behind from the constant procession of pilgrims.

I walked back up to the top of one of the ghats that gave good views of the river and the opposing shore, still farm and pasture land. I glanced out over the breadth of the river, which flowed so slowly it almost seemed to stand still. I then looked back towards the frenzy of activity at the river's edge. The burning never stops. It goes on day and night, year after year, and has for millennia. How many millions have been cremated and entered the river at this very spot? The essence of Hinduism was captured right here on the banks of the Ganges, as it moved slowly, relentlessly, eternally eastward. It represented nature in its purest form—the endless cycle of life and death.

Then I looked out beyond the black and green shore and all those little flowers and the hundreds of bathers and worshippers, beyond the boats that ferried tourists and travelers to and fro, out into the middle of the river, where I saw movement—four large gray shapes, four feet long, moving gracefully and steadily, emerging and submerging.

They were dolphins—Gangetic dolphins (*Platanista gangetica*)—one of the few freshwater dolphins on earth and one increasingly threatened by pollution here in India. But surprisingly this stretch of the Ganges was relatively clean, as most pollutants were organic. It is a health hazard to be sure—especially for the devout believers at the shore. But the dolphin survives, at least for now.

I made my way back to my hotel and came across a group of vendors and ramshackle food stands. Behind them on the disintegrating cement wall was a tattered picture of Shiva, sitting serenely on his tiger skin. An appropriate place, I thought, for the god of destruction.

Holy men, or sadhus, sat quietly a few yards away, clad only in loincloths. They were in the shade of an enormous peepal tree (*Ficus religiosa*) and did not move or even take note of me as I approached to take a closer look at it. The tree was 70 feet tall and about as wide, with smooth gray bark covering a gnarled trunk and giant curved limbs. The heart-shaped leaves, about the size of a man's hand, were thin and rigid, and they rustled with the slightest breeze.

The peepal is the great saxifrage of India. It is imposing and graceful. It is also useful. Bark is used for dyes, leaves for medicine, and fruit for food. But it is most famous as a symbol of holiness. For Buddhists it is the tree of enlightenment, but for Hindus it represents the entire spiritual world—its roots represent Brahma, the creator; its branches Vishnu, the preserver; and its leaves Shiva, the destroyer. The fruit represents the difference between body and soul. The body is like the skin, which, on the outside, feels and enjoys things, whereas the seed, on the inside, witness and reflects on things.

People have planted peepals throughout India for centuries, as reminders of the goodness and sacredness of life. People come the tree to pray or to tie red threads and pieces of cloth to signify special requests. Some only touch it on Saturdays to ensure good graces, and others see the killing of the peepal equivalent to killing a Brahmin. As I left I glanced back at the sadhus sitting below the heavy branches that blocked out the blazing sun, meditating, unperturbed, as they have for thousands of years.

Sadhus come from every station of life. Even wealthy men have been known to leave their money and families behind to live a life of poverty and prayer. In the past, many sadhus lived in remote forests, where they prayed and meditated in silence. They often came into contact with wild animals, especially tigers. And there are many cases of man-eaters killing unarmed sadhus, who did little to resist their fate as they were dedicated to a life of sacrifice, of which death was the culmination.

Hinduism is filled with many stories and legends extolling sacrifice, emphasizing human fragility and humility toward gods and nature, even wild animals. One of the most famous stories is actually a Buddhist myth. In one of the many versions of the Buddha's (the Prince Mahasattva's) life, he came upon a starving tigress and her cubs in the woods. Seeing her suffering he lay down and offered himself to the tigress, who devoured him and regained her strength, allowing her to save her cubs. The prince was then revealed to be the Buddha or "enlightened one." His bones were later buried and a shrine (stupa) was built there—in current-day Nepal—where pilgrims still come to pay homage.

But it was in Varanasi where the Buddha delivered his first sermons and gained converts—which is why it is often seen as the formal birthplace of Buddhism. Buddhism later spread throughout India and then into eastern and northern Asia. There are almost no Buddhists in India today, as Hindus drove them out or integrated Buddhist beliefs with Hinduism. Some believe this is another reason that Hinduism became more benevolent.

Likewise, Buddhism adopted many Hindu beliefs, especially the veneration of animals such as tigers. Followers of Buddhism are sometimes depicted riding tigers, as a sign of their ability to overcome evil. Buddhist practices also place many restrictions on killing animals and eating their flesh. This includes the eating of tiger flesh, once a widespread practice in much of Asia.

One of the most profound Buddhist influences on people's attitudes towards tigers was actually in China. Historically the tiger was associated with the yin—the forces of evil, whereas the dragon was associated with the yang—the forces of good. But as Buddhism spread through China, the tiger also became associated with goodness, which probably helped the cat survive until the coming of

Mao Zedong's communist regime, which, from the tiger's perspective, no doubt embodied a resurgent yin.

◆ ◆ ◆

After two days on the shores of the Ganges I left for Kolkata, which is located on the Hooghly River, a tributary of the Ganges and part of the vast delta where the Ganges finally ends its 1,500-mile journey and empties out into the Bay of Bengal.

This final leg took me through the poorest and most densely populated part of India—the states of Bihar and West Bengal. Historically these were flourishing centers of Hindu and Mogul culture, which again became important to the British. This area is where they established their foothold on the subcontinent, making Calcutta (then a collection of fishing villages) their capital in 1772. From here the British East India Company and later the British government (the British Raj) ruled India until 1911, when the capital was changed to New Delhi.

The climate here was wet and balmy, giving rise to lush tropical vegetation. The land was now flatter, the horizon wider, as we approached the coast. Rivers were more common and ponds numerous. Rice had replaced wheat as the dominant crop, and mangoes, bananas, and other fruit trees grew in profusion.

When the British came, they profited by taxing this intensive agrarian economy and by expanding cash crop production—mainly sugar, jute, and indigo. Calcutta's strategic location near the mouth of the Ganges made it a trading hub for goods like tea, timber, spices, and even opium that came from the interior. And as a naval base, this location allowed Britain to eventually oust the competing Dutch, French, and Portuguese from South Asia.

Calcutta and its environs made up the first Indian experience for many Britons. It was also where many saw their first tiger, as the cats were common in the marshy grasslands and lush forests that surrounded the city. It was not unusual to come to the edge of town for evening excursions and catch glimpses of the cat—even into the twentieth century.

The British became fascinated with India. Its "exotic" culture and strange plants and animals attracted the attention of many writers and artists in particular. The "picturesque" school of painting, with its romantic images of landscape, architecture, and animals, had deep roots in India. Painters were followed by photographers who shot pictures of ruins and people and, of course, many tigers, especially as dead hunting trophies.

Written accounts of explorations into the interior always included descriptions of tigers and other wild beasts, with hunting tales forming a staple of British writing on India. Scientists described tiger behavior and biology. And one of the most famous poems in the English language remains William Blake's "The Tyger." Blake never went to India, but the constant stream of stories and images from travelers allowed him to write one of the most evocative poems ever written about the animal.

The British can be credited for introducing tigers to the Western imagination, where they continue to have a strong presence. "Tiger" is the adjective given to dynamic growth economies, and tigers are common business and sports logos. Many plants and animals are named after the cat for their coloring or disposition. There are tiger snakes, tiger frogs, tiger salamanders, tiger moths, tiger beetles, tiger barbs, tiger sharks, and tiger lilies. No other animal, not even the mighty lion, evokes so much emulation in the Western imagination.

Yet Western views of the tiger were always paradoxical. The tiger was feared as much as it was revered and was always associated with evil or cowardice, a good reason to exterminate it or destroy its habitat. As a result, Europeans shot and killed more tigers than any other group of people who came to India, and in a shorter period of time. They also established the most comprehensive laws to protect them.

The British colonial government first enacted total bans on elephant and rhinoceros hunting, along with local and seasonal limits on tigers and other animals. They developed comprehensive forestry laws, and British big game hunters founded the world's largest wildlife conservation organization, the World Wide Fund for Nature (in East Africa), which later helped fund Project Tiger.

The scramble to protect nature was also in response to the revolution Britain unleashed in India. The British brought commerce and industry and modern politics. And the train was the engine of change, bringing economic growth and unifying the country politically. Trains also changed human relationship to nature.

Nature was exploited with an intensity not seen since the days of the great Hindu yagnas, but this time it was to appease the new gods of progress and power. Timber was cut and minerals extracted, swamps were drained, crops were planted, workers were moved and resettled, and cities were built. Hunters came to the most remote corners of the land. Naturally, the tiger retreated.

The rails were followed by roads and cars, electricity and chemicals. Cities grew and became more interconnected. Populations grew and required more resources. By the time of independence in 1947, the British had created institutions and infrastructure that would allow India to continue this development. Many Britons thought India would slide back into poverty and misery after their departure. But that was not the case. India continued to industrialize and is now one of the world's emerging economic powers.

The train had traveled several hours and it was now approaching dusk when a young man boarded and took a seat next to me. We soon began talking and he told me he was a university student, studying business.

"How do you like India?" he asked.

"I don't enjoy the cities very much," I told him. "But I like the countryside, especially wild places where tigers still survive."

I explained the dilemmas of conservation and how the tiger is greatly imperiled. I asked if he had ever seen a tiger.

"No. I have never been to any of these wild places. It is not a great interest of mine," he intoned. "I am more interested in people. India has 1.1 billion people and they must all be fed, clothed, given shelter and jobs."

He pointed to the window as we passed a congested intersection in a small town.

"You see, look at these people. They are poor. They have never seen a tiger and they do not have time to think of these things. They want to survive."

Then he smiled and pointed at me.

"You Westerners come here and want to save all these animals. It is a luxury. You don't have to think of survival." He shook his head in disapproval.

I asked him if it is an ethical duty to protect the cat. It is the symbol of India and has been a part of Hindu tradition for millennia.

"Wouldn't it be a supreme tragedy to lose the tiger in the land of the tiger?" I asked.

He shrugged and smiled. "I suppose you are right." Then he countered, somewhat indignantly, "But you Westerners destroyed your environment and now you tell us we should not do the same. You destroyed many forests and killed many animals when you came to India. Why shouldn't we? India must develop."

He paused and then assured me that the future was bright. India was progressing.

"Someday we will be like America and Europe!" he exclaimed with pride. "We will be just like you! Wait and see!"

He was so proud and confident that I did not have the heart to tell him that if 1.1 billion Indians lived like Americans the world's resources would quickly run out. And under such a scenario, there would be no place for tigers or any large creature that competes with people for food and land.

I changed the subject and we talked a little while longer and then readied ourselves for sleep and an early morning arrival at Kolkata.

I was awakened by the creaking of the train as we crept through the curvy tracks on our approach to Kolkata. It was still dark outside and all I could see were the small fires, surrounded by shadowing figures, in the hovels here on the outskirts of town. The burning wood and dung were mixed with heavy diesel fumes—the smell of the Indian city.

The young man was still sound asleep as we eased into the station and did not awaken until the doors opened and pandemonium erupted. I was up and out quickly and bid him goodbye as he groggily sat up in his bunk. I made my way onto the platform, now filled with hundreds of briskly moving bodies.

I soon found a chai wallah in a lighted area and sat down to have tea until dawn broke. I saw the student pass. I waved, but he did not see me as he disappeared into the crowd and then darkness. I sipped slowly, watching the constant movement around me, pondering the conversation from the night before. Here was Kolkata, the birthplace of modern India, now densely populated and rapidly industrializing. Could the old traditions still hold on in this rapidly changing country? Could they still be harnessed to save tigers and their forest homes, or would Western material values prevail and drive the tiger into extinction?

Some Indian conservationists argue the latter. Not only must India use its cultural and religious traditions to support conservation, it must go even further and reject Western values entirely. "Western pride, Western greed, Western materialism—that is what is killing India's forests and its tigers." I remember the old zamindar telling me this as I left Kalakad-Mundanthurai.

But was traditional India as harmonious as its defenders argue? The tiger was certainly an integral part of Indian consciousness and culture for millennia, and people successfully coexisted with tigers and other wild animals because of religious practices. But was this a deliberate decision or, as some argue, the result of (especially) Hindu beliefs that the physical world is an illusion? One simply accepts the authority of nature, tigers and all, even unto suffering and death. A world where people made few demands on nature benefited the tiger and its forest home.

But one contradiction of this fatalistic world is that it also accepted widespread human suffering. Few countries have such a history of human misery as India. This was still evident as I made my way through the streets of Kolkata to my hotel that morning. Poverty was openly displayed because it was considered natural, like wealth or beauty. Beggars joined sadhus on street corners, where they slept or waited for a charitable hand. I walked by one man without arms or legs, lying on a towel in the middle of a busy sidewalk, with a cup by his head where sympathetic passersby placed coins.

This fatalistic attitude also existed with respect to animals and nature. India's first prime minister, Jawaharlal Nehru, once remarked that nowhere is nature more revered than in India, and nowhere is it more mistreated. Animal suffering was also a part of fate.

I came across a dying dog later that afternoon on a busy street. It bled and trembled with pain, probably hit by a car. Hundreds of people walked past the suffering creature, but nobody took notice. I stopped and looked at it but felt helpless and soon went on my way, disturbed. I came later to see if it had died or been removed. It lay there, dead and covered with flies. Hundreds of people walked past it and nobody took notice. It was fate.

India's economic rise has been so violent, so rapid, so total that it too appears to be fate—something inevitable and irreversible. Would younger Indians, like the man I met in the train, who had never seen a tiger, simply accept this destructive system and its consequences as fate?

To return to the past and a simpler life where people and animals could coexist better is unlikely, especially for the majority of Indians, who no longer live near tigers and forests. The genie is now out of the bottle. Industry and technology and the rule of money are spreading and becoming universal—the so-called process of globalization. Going back to an older world where people and nature were in greater harmony, where tigers still roamed free and wild everywhere, is another romantic ideal.

But it is equally romantic to believe material progress is not only inevitable, but that it also has no price. This was the attitude Westerners had as they developed centuries ago, one that continues into the present and is now being adopted by people in developing countries, like the young student I met on the train. To live in a world completely divorced from nature is industrialization taken to its logical end. This romantic ideal will lead to hell on earth, and certainly the end of tigers and most other living things.

This is why cultural traditions are still important. They make sense of the world and channel human efforts. But they must now shift focus to harness and channel the new spirits of commerce and industry so that they do not lead India and its tigers to oblivion. Development must be compatible with, even an outgrowth of, a given culture rather than inimical to it.

As Indian history has shown, religious traditions and beliefs can create an environmental ethos. But economics can change, compromise, or even destroy these increasingly fragile customs. This is why conservation must also begin to address these new economic forces and channel them to protect tigers. Some even go further and suggest that conservation must be at the heart of a new economic paradigm—one that incorporates many traditional practices but is also keenly aware of the new social and economic world that is now emerging.

— 6 —

A Difficult Balance

The fusion of economics and conservation is nothing new. Conservation has always been about economizing scarce natural resources. What is new is a changing perspective on how resources are used and how resources (and nature in general) are valued. These ideas are the core of an emerging conservation paradigm commonly referred to as sustainable development or eco-development.

The fundamental goal of sustainability is to replace the idea of economic growth, where people benefit at the expense of nature, with one of balance, where people and nature both benefit. This implies a greater emphasis on economic efficiency as well as a long-term, rather than short-term, resource use horizon. The idea is that resources should be used prudently to make them available to future generations while protecting the future integrity of ecosystems.

Another focus of sustainability is political and economic. Sustainability emphasizes the devolution or sharing of power between the government, international organizations, and local communities. With respect to economics, importance is also placed on developing local markets and production rather than relying on national or global forces.

There is also a spiritual component to this paradigm, one that harkens back to the older religiosity and its reverence for nature. Nature is not just a storehouse of raw materials but also a source of beauty and mystery, a place of well-being, rest, and reflection. As such, conservation must be more than simply "saving" wild areas and wild animals from a hostile sea of humanity. Sustainability tries to create a new ethos, a new culture, in which wilderness and wild animals can once again be a part of human life.

Sustainability is a broad concept with many applications. As such it has attracted many critics. Many argue it is so vague that it can be applied to almost anything. Recently many business corporations have craftily added "sustainable" to their marketing lexicon and now use slogans like "sustainable growth" or "sustainable profits." Others argue it is another romantic ideal, too lofty to ever be put into practice, let alone become an alternative to conventional conservation and economic practices.

Yet supporters counter that eco-development is eminently realistic. It is a direct response to the rapid plundering of nature and resources, which leads to their reevaluation. Their scarcity is precisely what creates new conservation methods, even a consciousness shift regarding the human role in nature. Sustainability is the only *realistic* alternative to constant economic growth and consumption.

One type of place where sustainability may have the most success is at protected areas, places where ecosystems are still intact and where local communities still derive livelihoods from them. In India almost every tiger sanctuary has some type of eco-development program in place—alternative energy, improved agriculture, and ecotourism being examples. Most also have designated resource use zones—core areas for wildlife only and buffer zones where limited human activities like farming, timber harvesting, and grazing are allowed.

One project that is particularly instructive is Buxa Tiger Reserve in West Bengal. It is one of the most ecologically diverse in the country, with over 1,000 plant species and 400 vertebrates. It has also been exploited longer, more consistently, and more thoroughly than almost any other forest region in India. There are few places in India where the history of modern economic development and its destructive ecological and social impact can be so fully observed besides Buxa and the surrounding region. The area has also been fraught with ethnic tension and is perilously close to the East Asian poaching routes. All these factors make it a compelling case for sustainable development.

Buxa is in the duars, the section of the eastern Himalayan foothills that straddles the border between Bhutan and West Bengal. "Duar" means "door" or "passage" in Bengali, as these hills are the gateway to the east—a departure point for ancient trade routes to Tibet, Bhutan, and China.

Historically, the duars were remote and virtually uninhabited except by tribal peoples such as the Bodo and the Rava, who lived by hunting, gathering, and shifting cultivation. The Hindu and Muslim societies of the adjoining plains came periodically to cut timber, hunt, and capture elephants. The forests remained intact almost everywhere, and tigers were common.

With the arrival of the British in the 1700s, the forests were exploited in a more systematic manner to supply not only local demand (as had been done for centuries prior) but also the burgeoning national and international markets created by the British Empire. This early economy was a frontier economy, in which forest peoples were used for the hard work of clearing forests, planting crops, and building settlements. The ecosystem was drastically simplified.

Forest communities (mainly tribal peoples, but also some Hindu and Muslim farmers and herders) also changed as they were settled into "forest villages," part of the British "civilizing mission." This system gave the British greater economic and political control over them. Villagers were now part of the wage economy, and their income could be taxed, giving greater economic power to the colonial government.

The next phase of economic development in the duars was the expansion of the plantation economy—teak for timber, but most importantly, tea. The forest ecosystem, which had been altered through systematic timber harvesting, was now completely destroyed and replaced by these monocultures. The tea plantations also required huge amounts of labor, which had to be brought in from outside of the region. Hundreds of thousands of landless laborers came from central India, Nepal, and Bhutan.

This was an entirely new society—a rural proletariat bound to harvesting leaves in assembly-line fashion and then processing them in the adjoining tea factories. They were part of a growing wage economy that freed people from the vicissitudes of subsistence agriculture, only to make them dependent on the market. This new class of workers also had little knowledge of the forest. Like their British overlords they saw the forest as a commodity—for fuelwood and land to grow food.

The India-Pakistan War of 1971 brought a further flood of refugee-laborers from Bangladesh.[1] Today the tea-growing region of eastern India (which stretches from Darjeeling to Assam) supports a population in the millions that represents many ethnicities and religions. Poverty, crowding, and cultural differences have created numerous tensions that are readily exploited by unscrupulous politicians. In such a Machiavellian environment wildlife and forest conservation is a low social priority.

The most recent economic phase has been the expansion of mining, which coincided with the rapid urbanization, industrialization, and population growth of the post-independence era. The duars are rich in many important rocks and minerals like slate, quartzite, and especially dolomite, which is used to make cement—an essential material for any modern economy. As a result, the duars forests are now dotted with mines of varying sizes.

Mining is highly destructive, requiring armies of laborers who toil for years on end. They first strip away acres of forest, then remove all topsoil to reveal the desired rock, which is then excavated by explosives. The mineral-rich rubble is carried down steep mountain paths by foot to waiting lorries that ferry the material onward to manufacturing centers. It is a slow process but after decades can level entire mountains.

By the 1980s, this combination of economic forces, plus the usual farm and grazing pressures, had reduced the duars forests to a few fragments.[2] Buxa is the largest of these fragments at about 300 square miles, having a tiger population of about 15 cats with a few more roaming the reserve periphery. It is oddly shaped, with one main forest strip along the Bhutan border from which four thin, and not entirely contiguous, fingers extend southward. This distinctly gerrymandered shape was created to avoid the many surrounding villages, mines, and plantations. It is also the result of blind economic growth, which gave little thought to maintaining the integrity of the forests and their inhabitants.

Today the stated goal of the Buxa project is to protect these remaining forest tracts and its precious tigers. The unstated, and far more important, goal is how to stem the socioeconomic momentum of the last two and a half centuries to create an economy that can benefit people as well as forests and tigers.

After two days' travel from Kolkata, I arrived at the village of Alipurduar, where I stayed in the only tourist accommodations at Buxa, a colonial forestry rest house built back in 1870. It was a spacious building, painted turquoise, making it stand out in a wide forest clearing. It was two stories tall, with several rooms and an adjoining kitchen run by an old Nepalese cook nicknamed "Pampah." I was given room and board, including afternoon tea, for only a few rupees a day.

On the second day of my stay I was joined by a biologist from Kolkata, a Dr. Sutterjee who was doing plant surveys at Buxa to count native species and examine growth rates. We became fast friends and I accompanied him on his excursions during his three-day visit. Our first trip took us to the heart of some old growth forests. These are monsoon forests, similar to rainforests, with towering trees covered by vines and creepers, their crotches festooned with growths of ferns and orchids. But the canopies were not as dense, allowing herbs and shrubs to grow across the forest floor. Grasslands were also common, covering large areas around the many streams that flowed out of the higher elevations. The duars winters are also cooler and drier.

These were once some of the best tiger forests in India and, like the western terai, supported many large animals like elephants, rhinos, and water buffalo. This made the duars a popular hunting destination. The Buxa forests were among the best, which is why the maharaja of nearby Cooch Behar made it his private hunting preserve. He and many visiting dignitaries shot hundreds of cats there just a century ago.[3]

We parked at the edge of the forest where the road ended and walked down a shady path to a watchtower, a stone structure about 30 feet tall, once used by the forestry department but now open to visitors. I climbed up the stairs to the top, where I got a bird's-eye view of the woods, while Sutterjee walked up and down the forest paths that radiated out from the tower, busily observing and recording his findings. I watched the trails for over an hour hoping some animal would appear. Birds were active, but nothing more.

Sutterjee returned. "The number of plants here is incredible. And look at the size of these trees; some are hundreds of years old. This is one of the most diverse forests in India." Then he cast his arm about and lamented this was the last old forest left in all the duars. "These great forests once covered all these hills. But that was very long ago. All have been replaced by farms, especially big plantations. This whole area is dependent on the plantation and has been for a very long time."

He began to explain the timber and tea economies and then suggested we go out and see them firsthand. "You must see them to understand their impact and the contrast to this old forest."

We drove out of the old forest into open country, where we stopped at a wide nallah, about a hundred feet across, where a small stream, which swells to fill the entire bed in the monsoon, meandered through a sea of dry, pale cobbles. We walked across it to the stream's edge, where Sutterjee stopped and pointed to a block of woods that grew along the opposite shore behind a thick band of grass. The trees grew tall and straight, creating a uniform blanket of green covering at least a hundred acres.

"It looks very nice from the distance, doesn't it?" Sutterjee exclaimed with a hint of sarcasm. "But if you look at the ground you will see nothing grows there. These trees are planted close together to increase the yield, but they cannot support any vegetation, which means fewer deer and boar and that means fewer tigers."

Sutterjee told me that one-third of Buxa is covered by teak plantations. So in effect one-third of the reserve is unsuitable for wildlife.

"How many tigers could live here if all forests were returned to their natural state?" I asked.

"At least double. Maybe more—thirty to forty cats. Remember this region was once prime tiger habitat and they were the biggest and most beautiful cats in all of India—royal Bengal tigers!" Then he pointed back to the trees. "The funny thing is that teak is not even native to the duars; it is from south India. The British brought it here because it does well in this warm climate."[4]

Teak (*Tectona grandis*) grows fast, has few pests, and can be planted close together, making it a most profitable timber species. The Indian government has greatly expanded the number of plantations since independence.

"India is growing and we need timber," he added. "It is a difficult balance."

Sutturjee then took out his binoculars and gazed at distant hills about a mile away. "Here, look over at that mountain. You see there are no trees left. It is because of all the mining here—all dolomite."

An entire mountainside was removed, leaving behind an ashy-white scar, offset by patches of dark forest.

"No forest will grow there for decades, maybe centuries. And then during the monsoon, it erodes more, dumping mud into the rivers. One good thing is that many mines have been shut down since the project began. But the people don't like it. They need jobs. It is a real dilemma.

"But you know, tea is an even bigger problem. Most forests in the duars were cut for tea plantations. Like teak, the British brought tea here. It is not native to our country. It is from East Asia. But India now grows a lot of tea and it is very lucrative. It is our national obsession," he laughed. Then he turned and became pensive, "The problem is that the more tea we drink, the more plantations we need, which means we must cut more forests. And that is bad for wild animals. Tigers and tea do not mix.

"You see, tea only grows in a few places in India like the duars, places that have the right climate. [The Western Ghats are the other major tea-growing region.] The problem is that these perfect climates are also home to our most magnificent forests—like these here," he emphasized. "They are so high in diversity. And now all sacrificed for tea!"

Tea was next on our itinerary.

We left the reserve and came to an open landscape of rolling hills, thousands of square acres, all covered by bright green tea shrubs (a type of camellia) that were kept at a uniform height of three or four feet to facilitate harvesting. We stopped at the roadside in front of one, where I could see dozens of woman cutting the tender tips of the plants and placing them in big cloth bags straddled over their backs.

"You see the women do the harvesting and the men do the curing. Look up there." He pointed to an industrial-style building up the hill. "You see that smoke coming from the chimney. That is from an oven that cures the tea. And it uses wood. In the old days they just cut it out of the forest. But the forest is now gone so they grow timber in special plantations." I could see eucalyptus and other unidentifiable trees (all nonnative) planted in square plots all around the facility.

"Tea creates another problem," he said with a half smile. "You see, we Indians like our tea very very sweet." I nodded in full agreement. "And sugar plantations also take up a lot of land. India is one of the world's top sugar producers!" he said, exasperated. "So as demand for tea increases, so does demand for sugar, which means more land is needed."

All these factors are so interconnected, and all have an impact on the tiger. Here was a perfect example of how national and global economic forces have completely changed a local ecosystem. It is even more troubling because the consumers for these products are so disconnected from the consequences of their economic habits. One of the costs of a wealthy consumer society is that luxury goods like tea, sugar, coffee, cocoa, even opium and coca leaves, all grow in areas high in biodiversity. And as the global economy grows and more consumers demand these tropical treats, the pressure on forests and their inhabitants will only grow.

I asked Sutterjee about the people who work here. Where do they live? He shook his head. "I will show you. It is another problem."

So we drove a bit farther up the road to a makeshift village—a housing project for plantation workers—where we stopped. They were tiny concrete block buildings, built in perfect rows—just like the plantation they were built to serve. Many were dilapidated to the point of collapse and all overcrowded. Dogs and pigs roamed about garbage-strewn streets, watched by listless children and elders, who gathered around the front steps.

"Look at this place!" Sutterjee exclaimed. "Just like a Kolkata slum. Right here in the middle of the countryside. There are dozens of these shanties all over the duars. All because of tea! These people are poor, so they plunder the forest for wood, for their animals, for anything. They have little respect. They just want to survive. It is a dilemma here. Everywhere in the duars it is the same."

As we returned to our car I asked him if any of this could be changed to reduce the impact on the forests. Do the forest and the tiger have a chance?

"Tea and timber are so lucrative. It is hard to convince government or industry to replace these lands with wild forests. They would lose money and jobs. People desperately need jobs here. This is a poor place and there will be great anger if jobs are lost. The tiger will suffer."

"But what about all these teak plantations inside the reserve boundaries?" I asked. "Couldn't they be slowly cut and replanted with native forest and grasslands? It could create jobs and help increase the tiger population. Isn't that a possibility?"

He gave a contemplative nod. "Yes, yes, that is possible. But it will take time. You really have to change people's minds about tigers and conservation. People have to see it as something important." Then his eyes lit up. "You see this place has long been the land of timber and tea. It must become the land of timber, tea, and *tigers*. If that happens, there will be hope.

"Here is where tourism can play a role. If there were more attractions you could make money and create jobs from tourism." Then he paused. "But you must be realistic. Tourism will never be big here like at Kanha [in central India] or Corbett. Animals are just too hard to see in these forests and we are far from the main tourist routes. So we cannot rely on it."

Ecotourism is often touted as one of the sustainable programs that can bring money, create jobs, and protect habitat. But the problem is that tourism can also despoil, by bringing noise, pollution, and wealthy investors who monopolize the tourist trade. A bigger problem is that most Indian tiger sanctuaries do not have many tigers. And most places, they are rarely ever seen. So tourism is part of the sustainable package, but it cannot be seen as the only solution.

"So there are solutions," I emphasized.

"Yes," he agreed cautiously. Then he paused and turned, with a look of consternation. "But there is an old saying in this country—for every solution in India, there is a problem. It will take time."

After Sutterjee left for Kolkata, I met up with a group of young men—all from the surrounding area—who had formed their own environmental organization. Their goal was to educate local villagers about conservation, especially the importance of protecting tigers.

They came up to the rest house that evening as I sat out on the veranda. Word had got out that a visitor interested in tigers was in town. Five arrived, politely introduced themselves, and then sat down with me. I called Pampah to fetch some drinks. We began discussing the project and tigers, periodically interrupted by bats that flew through the wide veranda. Occasionally one would skid onto the smooth painted floor and just lie there, unable to get up. Apparently this was not uncommon as one of the men casually got up, grabbed the creature by both wings, and flung it over the railing. This happened several times throughout the evening.

The eldest of the group, maybe 25, was a handsome, serious man named Prabut. He was also the best English speaker and soon became the group spokesman and translator. He explained how they had all known each other since childhood and became interested in animals during high school.

"We had a very passionate teacher who would take us out into the forest and show us all these wonderful things. We have him to thank. So now we have taken it upon ourselves, as a duty to Buxa, to spread the word among our community. So we spend a lot of time just visiting people and discussing conservation."

Another man then interrupted and explained, in broken English, how they had helped with the Global Environmental Facility (GEF) project here.[5] "They had monies for farming and cattle," he explained. Prabut continued, "GEF was the big project here. It is international and they gave money to help with many projects, like farming, to reduce pressure on the forests. They have a lot of good ideas and we visited the projects often."

Then I asked about their money.

"Did you get anything from GEF or the government?" They collectively shook their heads.

"No, no. We receive no monies. We just do it because we are dedicated to Buxa. Maybe someday we will become big and famous and will get some help. Right now we all live with our families. We help them out and they are supportive. My father wonders why I don't just get a job," Prabut added. They all nodded in agreement and laughed.

The next day Prabut and two of his friends joined me and a driver to visit some of the GEF eco-development projects, located in a hamlet just on the outskirts of town. All were designed to reduce pressure on the forest and to use resources more efficiently, more sustainably.

We first came to a pisciculture operation—a pond about 100 feet square and 15 feet deep that was about half full. This is one of three dug here in this village to raise fish and ducks. We walked over to it and watched two boys cast their nets and pull in small fish. "This is a good project here, as we have a lot of rain; the ponds capture the water and then we can use it to raise fish and ducks. During the dry season it also helps with water so that people do not have to go to the forest to fetch it."

"Do people eat these fish locally?" I asked. "No, no, most here are strict vegetarian. But some do. Most people just dry the fish and sell them to local buyers who then take them to the towns and cities. These ponds are fairly productive—as long as the monsoon comes. And it always does. They cost little because the fish eat what's in the water."

Prabut then walked us into the village center to a hut where several old women were busy spinning cloth with a large weaving machine. "The project gave us this machine and now we use it to make cloth to sell or to sew into garments. The sewing goes on next door." He pointed to an adjacent hut filled with a dozen teenage girls busily assembling clothing.

We walked back outside as Prabut pontificated. "See all these projects can create jobs that are not dependent on the forest. They are trying to be efficient with resources. The project also helped fund small stores and helped with agriculture. But one of the biggest struggles is dealing with our cattle. How can we reduce the number of cattle that come to graze in the forests? That is priority. I will show you a program that is trying to do this."

We walked back past the fish ponds to an open field, where a newly built livestock pen stood. A young man came out and greeted us. He knew Prabut and they began talking. Prabut translated. "He has three cows, all enclosed in these stalls with a small adjoining pasture. The whole thing takes up less than half an acre. These are all hybrids that produce much more milk than Indian cows. See here, they are fed mainly grain, so they do not have to graze in the forest. It is an intensive operation. This man spends more time tending the cattle, but he does not have to follow them into the forest. So wild animals or cars do not kill them."

"So, is this the future of cattle in India?" I asked Prabut, who then turned and posed the question to the farmer. He smiled and nodded enthusiastically. "Yes, for him it is," Prabut said. "He is happy with this program. He makes money selling milk and is always close to his house, which is just right there." He pointed to a small brick house in the near distance.

Prabut interrogated further. "There are problems though. These hybrids are not as sturdy as our cattle, and feeding them is expensive. So it is not a total solution. The biggest problem is the culture. This program will never be accepted by all farmers. Many are so tied to the old ways. Not only are cows sacred, they are a sign of status. Even if they don't produce milk, people still want large herds!"

We thanked our host and returned to the car.

"You see," Prabut stressed, "we must try to convert one farmer at a time. Among the younger ones, who are more business minded, these programs can work. It is a good start, but it will take time. The cow is so central to life in India. One must be very careful. It is good to build upon the old ways. Improve them, but not destroy them. Otherwise there will be conflict."

Cattle and grazing is another conservation problem throughout India, as cattle compete directly with wild game for fodder. The impact on the tiger is indirect, because more cattle means fewer deer, which means fewer tigers. India may have as many as one billion livestock—cows, buffaloes, sheep, and goats—more than any place on earth. Even on the drive back to town we must have passed over a hundred cows, all foraging in field and forest. Almost nowhere are cattle enclosed in pens. They just roam free.

There is no tiger sanctuary in India that is free of cattle. Eliminating them completely is unlikely, but reducing their impact is possible. Raising hybrids and enclosing cattle is one way to reduce the impact. Some projects have started paying farmers to castrate older bulls to reduce reproduction rates, while others have worked with farmers to graze cattle only in designated areas. It requires great patience and creativity, but it is of utmost importance.

◆ ◆ ◆

Prabut and I went out alone the next morning to visit two more villages at the periphery of the sanctuary. The first was about five miles away, on the same riverbed I visited with Sutterjee a few days earlier—but farther upstream.

The drive took us back through dense forest. Sal was common here but now a minority among many other trees—both great and small. One reason the British planted teak was that these existing forests, while rich in diversity, were poor in timber species. So they replaced them with productive plantations, rather than allow the native forest to regenerate, as was the case in the western terai, where sal is dominant.

The narrow road soon became a dirt track that took us into a low-lying area, moist and prone to flooding and giving rise to impenetrable masses of vines and thickets. Prabut leaned over and spoke to the driver and then turned to me. "I told him to drive steady and not to stop for anyone on the roadside. You see bandits sometimes frequent this place."

"When was the last attack?" I asked pensively.

"Maybe two years ago. It has slowed down. But I am always cautious when I visit these forest villages. Young men resort to banditry because there are no jobs in these places. They are far out, so they have not benefited much from the government or GEF programs."

A forest clearing, home to a small village of a few hundred people, came into view. We drove up and parked in front of the forester's office. We walked up to see him, but he was already out. Prabut then pointed to the river. "Let us walk out here to the river; I will show you some of the problems here." We were soon joined by a group of young men; Prabut knew some, and they began talking. By the time we reached the riverbed we had an entourage of about 30.

I could now see the dolomite mine up close; the gaping hole almost shone in the full sun. "See that is the dolomite mine, where these men once worked. It is now shut down," Prabut explained. I looked out into the crowd of sullen faces, eyes fixed on us. A few started to prod Prabut, and then more voices joined in with an air of complaint. Prabut tried to remain calm.

He turned and explained to me that these men know him, but they still do not fully trust him. "Every time I come here they confront me like this and they think I am with GEF or Project Tiger and want to know why nothing has been done here. They have lost jobs. Even here in the streambed, they used to make money collecting boulders for construction companies. But even that has been denied them. I try to explain to them that it's for the good of conservation. But they have no interest in tigers. They need their work. But, I keep coming back, hoping to convince them. But I need to deliver good news to them.

"Look, shutting the mine was a good idea. But the project should allow them to continue collecting boulders. Even the forester here says it does little damage. Boulders come down from the mountain with every monsoon. It would be a good compromise. I have talked to the forester about it and he agrees with me, but everything moves so slowly with the government. The people don't want to wait. Look how angry they are!"

Prabut eventually calmed them down. "I think it is best we go now." He looked at me, exasperated, sweat running down the side of his face. We walked slowly back to our jeep, and Prabut signaled to the driver to start the engine. We were still being followed by the men, now silent, but still sullen. We too remained silent, talking under our breaths, fearing a loud voice or sudden gesture could ignite them. Both of us slid into the jeep and we departed quickly down the dirt rack. Neither of us looked back.

◆ ◆ ◆

Many conservationists see the GEF project and other international projects as the tiger's salvation. They are seen as signs that the global economy also has a green side that can help bolster conservation efforts and work with local communities to reduce their impact on the environment. At Buxa new ideas have been introduced and projects begun. People are interested in helping conserve. But other problems have emerged here and elsewhere.

One problem has been project goals. They are lofty, creating high expectations among project staff and populace alike. While sustainable programs have been implemented, the effects have been uneven as the projects have only been able to reach a small part of the population. This has created resentment, as we witnessed at the river village. People were promised work that never materialized; others were denied work.

There is also a problem of suitability. International organizations have little knowledge about the unique cultural, political, economic, and ecological factors that exist at a given tiger sanctuary—and they differ everywhere. General models are developed and expected to fit every circumstance.

An even bigger problem has been money. Large sums were spent at every project site—sometimes 10 times the normal operating budget. This windfall leads to abuse. A large percentage of project funds are consumed by inflated salaries, consultants' fees, and new and advanced technologies; given to nongovernmental organizations (not all of which are legitimate); or simply stolen. Money has attracted criminals and in some cases actually found its way into poachers' hands to kill tigers—an investment that could make even more money!

Money was also allocated in a short time frame. The transition to more stable and sustainable economies will take decades. A 5- or 10-year project is not enough time. Allocating smaller amounts of money and spreading this over a longer time frame is a better approach.

The source of money may be the biggest problem. The very global actors, especially the World Bank, who are promoting sustainability are also aggressively promoting mining, timber, and massive agricultural schemes that are having a devastating impact on the environment and on local communities the world over. Their much-vaunted green initiatives seem like nothing more than public relations ploys, covering up business as usual.

Does this mean that any and all eco-development initiatives are worthless? Not necessarily. The problem is not the concept, but organization and leadership. What really need to emerge are authentic local initiatives, organized, directed, and staffed by local people. Such groups could still receive international or national assistance (especially from a better organized and led Project Tiger), but they would also have a level of autonomy to be better able to control their agenda and resources. Environment movements must begin at the level of the community and then radiate outward. They must, as all living things, be rooted in the ground—growing bottom up rather than top down.

Some will argue that such efforts are too small, too disorganized, too piecemeal to be successful. Tiger conservation is a big problem that requires big solutions, and only big institutions can do this. But if the Buxa case reveals anything, it is that these big bureaucracies have had terrible environmental and social records. If that is the precedent, local groups can fare no worse.

My final visit with Prabut took us to a hill village near the ruins of the old British fortress that once guarded the frontier. We drove 10 miles through thick forest, then through open land and teak plantations, over several streams, through villages to the very base of the foothills, where the road ended abruptly and became a footpath, well worn by cows and people.

We readied ourselves for a two-hour trek, leaving our driver behind, who gladly took the opportunity to lie down and take a nap in the back of the jeep. Prabut led the way up the path, which rose steeply and then disappeared in the woods. "In the rainy season, this place is crawling with bugs and leeches," Prabut told me as we moved through heavy brush. "They fall right off these plants and get you!" But now in the cool, dry winter, we marched on unperturbed. Many forest trees were leafless, giving us views of the forest floor and glimpses of the deep ravines that straddled either side of our hill.

At one point we reached a wide opening where we could see for miles. It was a vision of primordial India—velvety forest covering several hills that rolled gently toward the low ground, where a mist-covered river wound off into the distance. The only human presence was a few tiny villages and their square farm plots far in the distance.

We climbed higher, up steep terrain, through more dense woods and then into flatter terrain, which, after 200 yards, opened up to a wide summit. I could see a hamlet in the distance, surrounded by pastures and farmland.

Our first stop was the old fortress, which was to the right on a hillock overlooking a series of hills and valleys that separated India from Bhutan. The fort once guarded the boundary and was later used to house prisoners. When they achieved independence the Indians ceremoniously destroyed it, leaving a few walls and rooms behind, now covered by vines and shrubs. It was not much to see but was an important icon for Indian independence. The old mountain trails that once served as main trade arteries to the east had largely fallen into disuse, now used mainly by local villagers and smugglers who move weapons, drugs, and tiger parts to lucrative destinations farther east.

We walked back towards the village to a tea stall where a dozen men, of varying ages, sat talking. Prabut knew a few and he introduced me, and we began talking about tigers. I was curious if anyone had seen any here recently in this more remote part of the sanctuary. No one had, but a few men had seen a cat in their lifetime, some as recently as five years ago. All agreed they were still around but were very secretive because of all the people here.

Then the oldest man in the place, probably 90, came over to Prabut and starting talking to us in a low voice. Soon everybody quieted down and listened. Prabut translated to me.

"He said he was once a government tiger hunter, paid to kill man-eaters and cattle lifters—one of only two men in all of West Bengal with that special license. He shot 32 tigers between 1957 to 1962, when he finally put up his gun and retired from hunting after his wife begged him to change careers."

The old man went on to tell us he had many dangerous hunting experiences; the most harrowing was in 1957 when he was summoned to a forest village to kill a cattle lifter. A group of men ran from their village to his and descended upon his hut to beg help. The cat had become increasingly aggressive and would not leave the village. When he arrived, the cat was still prowling about the huts, whose tightly shut doors were packed with terrorized inhabitants.

The tiger immediately spotted the hunter and charged, allowing him only a hasty shot that grazed its mouth. The cat became enraged, knocked him to the ground and beginning a vicious attack that went on for 30 minutes.

He fought alone, hitting the animal with his fists and rifle butt and trying to keep it from inflicting a serious wound. But he grew tired, allowing the animal to bite him in his left arm.

The old man paused and then looked at us as he rolled up his sleeve to show us the fang marks. There were two deep puncture wounds, which had left his hand slightly disabled. Then he pulled up his shirt and showed several claw marks, now thin lines, barely visible.

He remembers the unbearable pain. Worse yet, he was losing a lot of blood because the cat would not let go of his arm. In a last desperate move he groped for his bayonet with his free right hand and managed to grasp it. He took a deep breath, plunged it in the tiger's side with all his might, and hit directly in its heart. The tiger died instantly, let go of his arm, and slumped to his side. He got up and stabbed it repeatedly until he was sure it was dead.

The man lay down, exhausted, in a pool of blood, a mixture of his and the tiger's. The villagers finally rushed out of their huts and bound his wounds and then carried him to the hospital. He lost a great deal of blood, causing him to faint by the time he reached the doctors. His wounds required over 100 stitches and many weeks in bed.

The old man then went silent and looked down at the ground. I asked Prabut to ask him what he thought of tigers after his near-death experience. He looked up and looked me in the eye, as if to convey his thoughts directly without speaking. Then he turned to Prabut and explained that despite almost dying and his slight handicap, he never hated tigers, not even man-eaters. They were a part of forest life.

"Today he feels great sadness that the forest and its animals are gone," Prabut added. "In those days, everyone knew and respected the tiger. He says this is no longer the case. The younger generations have no respect or knowledge of the forest. They want the forest for its wealth. He wished the forest still belonged the tiger."

As we left the village later that afternoon I asked Prabut whether the old man was right—do young people simply want the forest for its wealth. He paused and reflected and, in his usual somber way, nodded his head. "Yes, in general he is right. I am sorry to say."

He paused and then turned to me and said, with stiff upper lip and determined voice, "But, there is also hope. Many of my generation know things are changing rapidly. We cannot ignore that the forest and tiger are endangered. We are committed to help. We change people's minds every day. It is a lifelong struggle. We can do it. We must do it."

— 7 —

Tigers and Tribes

Some conservationists believe the most sustainable relationship between nature and people is found among India's tribal peoples. Their low populations and simple technologies have ensured a relatively harmonious relationship with tigers for millennia.

However, this ancient relationship is becoming undone as populations increase and forests shrink. As a result, tribal people want greater control over these remaining forests and their resources, including greater cultural and political autonomy from the government that now controls them. Some are turning to violence to stake their claims—targeting government personnel and property. Others have organized armed resistance against the Indian army and police.

Wherever violence erupts, wildlife is a casualty. Militant tribal groups often work with poaching rings and crime syndicates to harvest the meat, skin, bones, and horns of wild animals to fund their operations. Wildlife is sometimes wantonly destroyed simply because it is seen as a symbol of state power.

At Nagarhole National Park in southern India, tribal people have burned down hundreds of acres of forest to protest against the government. In the state of Jharkhand in central India, tribal people have taken over numerous wild sanctuaries, effectively shutting them down, making forests and wildlife subject to their control. Eastern India, with its large concentration of tribal peoples, has experienced some of the worst violence.

The Assamese, really a nation of several million, have consistently attacked government personnel and innocent civilians, as well as plundered forest resources in the name of independence. Poaching has been particularly rampant at Kaziranga National Park, located in the vast grasslands of the Brahmaputra River valley. It is the last significant parcel of once-extensive flood plain habitat remaining in India and home to the world's largest one-horned rhino populations as well as tigers and other rare animals.

Some of the worst offenses have occurred at Manas Tiger Reserve in the nearby Himalayan foothills. It is a premier tiger habitat and a center of biodiversity. In the height of their assault in the late 1990s, rebels managed to control most of the park, during which time dozens of elephants and the majority of the rhino population was killed off. No one knows how many tigers and other species were lost.

Until conflicts between tribal people and the Indian government are resolved, the future of the tiger will remain in jeopardy wherever tigers and tribes coexist, because almost every tiger forest in India is also home to tribal peoples. To better understand these conflicts requires some background on tribal culture and history.

◆ ◆ ◆

Tribal people number over 80 million in India and are defined as a group because of their shared history as hunter-gatherers and shifting cultivators. Aside from these economic similarities, they represent a wide range of languages, physical features, and religious beliefs. They are also grouped together because they migrated to India prior to agrarian peoples like the Aryans and Moguls. But even these origins are diverse, as there were several waves of human migration that lasted for tens of millennia.

Anthropologists believe the very first were the so-called Negritos, a group of people with dark skin, curly hair, and short stature—similar to the African pygmies—but who probably originated in New Guinea.[1] These peoples were later displaced by or they intermarried with waves of hunting-gathering peoples from Africa, the Middle East, and Asia.

Even today these influences can be seen. Tribal people from central and southern India are darker skinned, often with curly hair revealing African or Middle Eastern ancestry. Those in eastern India have distinct Asiatic features. The greatest concentration of tribals is in central and eastern India. But they live in almost every Indian state where forests and mountains are found.

Tribal culture varies greatly according to climate and geography. Some tribal peoples, like the Kharia of eastern India, lived deep in the forests in small, isolated bands. They never developed weapons or tools because they existed almost exclusively by gathering forest produce. They wore little clothing and built makeshift huts well adapted to their mobile lifestyle.

Others, such as the Gonds of central India, were hunters and shifting cultivators who established extensive kingdoms centuries ago with relatively complex political and social organization. This is why the fabled prehistoric landmass that once included India, South America, Africa, Antarctica, and Australia, before it split up some 200 million years ago, was named "Gondwanaland."[2]

Some, like the Bhils of western India, were peaceful and coexisted, even intermarried, with their Hindu neighbors. Others were warlike. The Naga of northeastern India were some of the most notorious. They were headhunters who prominently displayed the skulls of their victims, which often numbered in the hundreds, on the walls of special "skull houses" or *morung*.

Whether hunters or farmers, peaceful or violent, all tribals eventually suffered the same fate with the expansion of Hindu, Muslim, and British civilizations. Like the tiger, they were pushed into the most remote and rugged terrain.

Historically, the Hindus considered tribal peoples primitive and uncivilized. But they were not part of the caste system and so had a higher social status than many low caste Hindus like the *dalits* (untouchables). For the most part, Hindus and tribal people lived apart. Although considered culturally inferior, tribal people were often granted political semiautonomy and became useful allies during times of war.

Tribal life began to change with the coming of the Moguls, who saw tribal beliefs and practices as heathen and subject to persecution. These Muslim rulers also created a more centralized and comprehensive feudal system. The system required tribal people to produce revenues and pay taxes. This forced them to slowly abandon their traditional economic life for settled agriculture.

The British accelerated this process by laying claim to all land under their rule. Farms and forests alike were under their control, usually administered through a government-appointed native landlord (zamindar). To satisfy the demands of their global economic empire, the British demanded ever greater productivity from these lands—they were converted to mines and plantations, which required huge numbers of laborers. Tribal people were drawn into this economy, joined by countless immigrants from all over the country (as the Buxa case well illustrates). Tribal people lost their ancestral lands and even became minorities there. Wildlife and forest conservation policies further deprived tribal peoples of living, hunting, and gathering on their ancestral lands.

The British did recognize some of these problems and passed laws that tried to protect the sociocultural identity of tribal peoples and to provide them with some political and cultural autonomy. Together with low caste Hindus, tribal people were collectively defined as scheduled castes.

After independence the Indian government centralized power and pursued policies of rapid industrialization that put renewed pressures on forests and natural resources. Despite some legal attempts to help tribal areas develop, those efforts were outweighed by continued economic privation and a loss of political and cultural autonomy. These factors finally drove tribal peoples to demand greater rights and reversal of centuries of discrimination and displacement.

One of the largest concentrations of tribal peoples in India is in the state of Orissa, in eastern India, where over a quarter of the population is tribal. Most live in the two mountain ranges that define the state's geography—the Eastern Ghats, which run along the eastern Indian coast, and the Chota Nagpur Plateau, which lies just north of it.

Proximity to the coast has created a balmy climate throughout Orissa that supports luxuriant vegetation similar to that found in the Western Ghats. Yet these are sal forests, which unlike the rainforests of the Western Ghats can support large herbivore and tiger populations. Orissa is also an important elephant and primate sanctuary.

The premier tiger reserve in these mountains is Simlipal, located in the northern part of Orissa, where the Eastern Ghats and the Chota Nagpur Plateau converge. It is 1,200 square miles in size, making it one of India's largest wildlife sanctuaries, and it has 80 to 100 tigers, also one of the country's largest populations. And it is home to many tribes, including the Kharias, Santhals, Mankdias, Mundas, Bathudis, Gonds, and Hos.

Simlipal is famous for its scenery, which is defined by ancient volcanic activity. Massive craters, mesas, and stacks of boulders punctuate the forested landscape—the largest expanse of sal forest remaining in India. But tigers are rarely seen.

I asked my driver Rabindra (who was also my cook and guide on this junket to Simlipal) whether he sees them often. He shook his head and held up three fingers. "Three times," he said. And he had worked here for almost a decade. "Most are deep in the forests, on the other side of the forest where people are less."

Rabindra was a young man with wife and child who worked sporadically as a driver in the nearby town of Baripada. He told me these park trips were his favorite job as they paid well and allowed him time in the forests. And because he spoke decent English, his services were always in demand.

The entry side of the park did have numerous villages; their mud huts were built of red clay with brown thatched roofs and stood out in small forest clearings. And they were all tribal communities. I first noticed the large number of tribal people in this region on the narrow pitted highway that took us from the town of Balasore, about 40 miles from Baripada and the park. All along the road there were small groups of women weaving together thin strands of jute, about 30 feet long, to make rope. From the distance they appeared like the average Hindu peasant, wearing saris, their hair in long neat braids. But upon closer inspection I could clearly see they were different. They had round faces, darker skin, and shorter stature. They worked stoically, never looking up, as we sped by.

Tribal communities here, as in many parts of India, have adopted not only the dress but also the language and many economic activities from their Hindu neighbors, like rope making, selling crafts, grazing, and farming. But they still lived apart from Hindu society. Many retain their traditional animist religious beliefs, and they still rely on hunting and gathering to survive.

As we approached the park entrance we encountered more tribal hamlets. But they grew thinner and the forests thicker once we entered the park. The terrain became hilly. Our jeep lumbered slowly up narrow dirt tracks that wound their way up the mountainsides, straddled on either side by tall stands of sal trees.

It was only mid-spring, but it already approached 90 degrees by midmorning. The heat intensified everything. It drew the aroma out of wild fruits, flowers, and frankincense trees. And the strong glare made the trees stand out in sharp relief against the deep blue sky above and the carpet of fallen yellow and brown leaves below.

Water was still abundant in these hills despite the months of rainless weather. Small streams flowed through bamboo-covered ravines, and clear pools gathered along the base of granite hillsides.

At one point we came to a small pool, fed by a tiny trickle from the rocks above, directly next to the road. We stopped to look for life—maybe a frog or insect. But nothing stirred. My guide went down on his knees and washed his hands and face and then dunked his head in the cool water, emerging with a sigh of relief.

"It will be hot today. Nothing moves when it is like this. The forest will be quiet."

Quiet except for one lone voice.

And we soon heard it—a piercing call that sounded like "Brainfever, brainfever, brainfever . . . " It increased in volume, reached a shrill, hysterical frenzy, and then dropped off. After a few moments it would start again. And it continued unabated. It was none other than the brain-fever bird, one of the most notable sounds of the Indian forest. As soon as we passed one bird we would encounter another. By afternoon they sounded from almost every corner of the forest—but I never caught a glimpse of one. They were always hidden away in the foliage of some tall tree.

Despite its piercing call, *Cuculus varius* is only the size of a pigeon. It is a relative of the cuckoo, and the two are sometimes heard calling simultaneously. And during the dry season they are often the only sounds in the middle of the day. The calls go on from midmorning until evening, sometimes all night during a full moon. They finally end with the coming of the rains, which here were still weeks off.

Rabindra slowed as we came to a clearing, where he pointed to the path that cut across the road. "This is an elephant trail," he remarked. "In the mornings and evening you can sometimes see them here." I looked down at the trail, well trodden and marked by dozens of plate-sized footprints. "It is hot. They are now resting deep in those woods." He pointed to a tunnel-like entrance into thick vegetation, clearly forged by the elephants, where the trail disappeared into darkness. Rabindra drove quickly by the entrance, perhaps in anticipation that some huge tusker might burst out at any moment.

We drove on another hour and were approaching the forest rest house, where we were to spend the night, when Rabindra slammed on the brakes. "Snake!" he shouted, pointing to the bushes a few yards ahead, into which a long, dark shape disappeared. "Cobra!" he said confidently as he looked me in the eye. "Big one."

It was the king cobra or hamadryad (*Ophiophagus hannah*), which can reach 18 feet—the biggest poisonous snake in the world. And when a full-grown specimen rears up and spreads its hood it stands six feet tall, making it one of nature's most frightening displays. This one was only a "small" six feet and fortunately long gone.

Despite its size and ferocity, the king cobra is not the one preferred by Indian snake charmers. They like the smaller Indian or spectacled cobra (*Naja naja*), which can easily be carried around in baskets, allowing the charmers to move around to ply their trade. Its spectacled hood looks like a pair of eyes, making it appear even more ominous. Although threatening, cobras are not the most dangerous Indian snakes. That distinction goes to the vipers, whose long rigid fangs can easily penetrate clothing and even thin shoes.[3] In contrast, cobras have short fangs and they must chew their venom slowly into their victims, making them easier to deflect.

But both kill tens of thousands of Indians every year—far more than the deaths due to all other wild animals combined. Most victims are peasants and forest dwellers, like the people who live here. And when they are bitten, many are unable to reach medical facilities and will die a slow and painful death. In some areas people have tried to eradicate poisonous snakes, but the rodent population then explodes, destroying crops and spreading disease. Poisonous snakes are simply a part of Indian rural life, necessary to nature's balance.

We soon rolled up to the rest house, a simple wooden structure once used by the forestry department, which stood in a grove of trees with open fields close by. These openings were used decades ago by the Maharaja of Baripada to hunt tigers and other game. The hunting stand, built of solid granite with arched gun ports, still stood strong at the top of a gently sloped hill.

I walked over to it as Rabindra prepared our camp and sat in one of the wide-open ports, where guns were once placed. Nothing stirred in heat of the afternoon. The only sound was a chorus of brain-fever birds off in the distance.

I returned to the meadow at dusk and soon spotted four small deer—barking deer (*Muntiacus muntjak*) or kakar, known for their curious dog-like bark. A pair of peacocks joined them after some time and all ate peacefully at the forest's edge as evening approached. And the brain-fever birds kept up their monotonous calls until it was dark.

After a few days at the rest house we moved on to the Barheipani Waterfall, which cascades 1,200 feet into a deep gorge. Our trip first took us into cleared forest and more tribal hamlets. We stopped at one tucked into a shady grove where a few women sat against the cool mud walls of their huts, tending little children. A lone cow grazed off in the open fields, surrounded by a blanket of shimmering hot air.

The women had baskets of forest fruits—astringent *jamun*—at their feet. Rabindra paid them a few rupees for some, which they politely accepted. He spoke to them for a while and then we went on our way.

"They are Santhal people," Rabindra intoned as we drove off. "This whole valley is Santhal people. Every valley has a different tribe." He started pointing to far-off mountain ranges and listing their tribal inhabitants. "We will meet some of these, except the Kharia. They stay deep in the forest. They are true forest people."

We soon came to another Santhal hamlet in a forest clearing that spread for hundreds of acres. There were low rolling hills throughout, covered with patches of dried grass, much of it overgrazed. Deep gullys formed all along the steeper slopes, where heavy rains had shorn off the last patches of topsoil to reveal the bright red soils beneath. They will deepen and widen with every passing year, exposing ever more oxidized infertile soils. The only thing these soils are good for is brick making.

As we approached the cluster of huts, which was near a wide gully, there were 20 people doing just that. Small children, young adults, even old women dug up the dirt, hauled it to a muddy water hole, and made a thick mixture. The batter was then poured into crude wooden forms that were placed out in the searing heat. They dried within a day and made good bricks—lightweight, solid, and weather resistant. They are sometimes sold in surrounding areas. It is one of the few economic activities here, but it ceases when the rains come.

During the rains many tribal people plant these fields. Historically they practiced shifting cultivation. This traditional method of farming is also known as "slash and burn" agriculture, because it required forests to first be burned or cut down. The cleared land was then planted for several years, until soils became exhausted. Farmers would then shift to another plot and repeat the process.

In the past, shifting cultivation was not detrimental to the forest as human populations were low and forests plentiful. Abandoned land would regenerate, and the new growth provided important fodder for elephants and herbivores. But with limited forests, the process is becoming too destructive. Most tribal communities now practice some type of settled or permanent agriculture, which relies on constant application of manure and labor to maintain production.

The economy is a major problem for tribal people everywhere in India. At Simlipal it is particularly acute, as there are few sources of permanent employment. There is seasonal work like brick making and timber harvesting. But there are no mines or plantations and almost no tourism to create permanent employment. This has obvious environmental benefits, but also costs. Even when forests are not cleared for plantations or mines, they are still exploited for hunting and gathering.

Tribal hunting has become a major controversy at Simlipal. The park is famous (or infamous) for the age-old ritual of *akhand shikar*, or the "non-stop hunt," named so because it extends throughout the entire dry season. All the different tribal groups carry on the ritual in their section of forest, except a few like the Kharia, who subsist almost exclusively from forest produce.

In these hunts, hundreds of men arm themselves with axes, clubs, spears, and poisonous arrows and bows and form lines that stretch a half mile or more. The underbrush is then set on fire, forcing animals to flee into the open where they are systematically slaughtered. Most are deer, birds, and small mammals. But elephants, leopards, bears, and tigers have all been taken. The real impact on tigers is the disturbance of habitat and the loss of prey species.

Park officials have tried to monitor the event, but the vast territory and relatively few game guards has made this difficult. The outnumbered guards are also threatened, attacked, even killed. In the past, these hunts did not have such an ecological impact as wildlife was plentiful and people were not. Now it is the reverse. It is another example of how difficult it is to overcome the momentum of tradition.

But that tradition has also changed. In recent years poaching gangs have used the traditional event as an umbrella for illegal hunting, including tigers and elephants. Organized crime also pays tribal people to illegally cut timber, which can easily be moved with improved roads and communications. Tribal people, because of their keen knowledge of the forest, are also hired to collect plants and animals for the food and pet trade. They are paid a pittance for items that fetch hundreds, even thousands of dollars in the global marketplace.

One animal seriously impacted at Simlipal is the beautiful hill mynah (*Gracula religiosa*). It is a favorite in the global pet trade because of its ability to mimic the human voice and its striking looks—glossy black feathers with white wing patches, yellow feet and bill, and conspicuous orange eye patches. Simlipal was once known for the largest population in India, and flocks numbering 20 or 30 birds were commonly seen. Today a mating pair is a rare sight.

And when a pair does produce chicks, many never grow to adulthood in the forest. Poachers steal the chicks from nests or trap juveniles (with glue or snares). The birds are kept in cages and domesticated and, when they are big enough, sent off to city vendors who trade them worldwide.

Frogs are another favorite commodity, sought out for their meaty hind legs. Orissa, because of its balmy climate and abundant water, is a frog haven, especially for the large species so desired in Southeast Asian and Chinese cuisine.[4] This demand has seriously depleted eastern India's frog population, to the point that many are now on the endangered species list. Sadly, controlling the killing and shipment of these smaller animals and plants is even more difficult than for bigger animals.

The plight of frogs, birds, fish, and plants seems insignificant when compared to that of tigers or elephants. But the loss of these microfauna could have an even greater impact on the overall functioning of complex tropical ecosystems than the loss of bigger animals. This is because these microfauna, which number in the millions, form the food base of the entire ecosystem. And when that base is undermined the entire system can collapse.

As we left the brick makers I wondered what they would do with the coming of the rains. Would they work for the poachers and gangsters who are stripping India's forests clean to satisfy a burgeoning global economy?

◆ ◆ ◆

One of the major tribal issues at Simlipal, as in most Indian wildlife sanctuaries, is village resettlement. There are dozens of tribal villages inside the reserve boundaries that put ever greater pressures on the forest and its resources.

The government resettled some communities living inside Simlipal's core area to its boundaries. But moving peoples who rely heavily on hunting, gathering, and shifting cultivation for survival is problematic, because many lack the work skills and resources to survive outside of the forests. Many simply wander back to hunt and gather in secret.

The government tries to compensate by providing housing, new land, and intermittent work. To the government this tradeoff is necessary for conservation to succeed. But to tribal leaders and their supporters, it represents a continued loss of tribal culture, creating a population wholly dependent on the state for its very existence.

Tribes such as the Kharia, who have been resettled or forbidden to engage in their accustomed way of life, have suffered the most. For them, every hill range in Simlipal is under the charge of a hill goddess who is responsible for its growth and maintenance. These goddesses answer to Badam, the forest god for all of Simlipal. Badam provides a generous natural bounty to people who pray and follow the proper rituals, which are closely connected to Kharia economic activities. But with government restrictions and the encroachment on traditional Kharia haunts by outsiders, this elaborate cycle of rituals and related sacred events has been severely curtailed.

For displaced Kharia, this severance from their forest and gods has caused untold grief. Many believe they have abdicated their responsibilities to their deities and as guardians of the forest and will be punished, causing natural disasters in the forest and famine among the people. Oddly enough, many would say that this is exactly what the tribal people are now experiencing with their removal from the forest and their changing attitude and relationship to it.

The question over tribal peoples has opened up a heated debate within the global conservation community and brings into question the viability of sustainable development. The idea of "wilderness" is at the center of this debate. It denotes, among other things, habitats free from *all* human influence, as close to a "pristine state of nature" as possible.

The wilderness idea is of Western origin, and it has been most faithfully adhered to by Americans and the British, both of whom forcefully removed indigenous peoples from large expanses of land to make way for wildlife sanctuaries (at home and in their colonies, respectively).

Moreover, it is a legacy that continues to shape environmental policy-making in postcolonial regimes like India. This "orthodox" view of conservation has been heavily influenced by the natural sciences, especially botany and zoology, with their exclusive focus on the nonhuman inhabitants of a given biome. And if the goal of conservation is to protect wilderness, then an important part of the conservation mission must be to rid wilderness of any human influences, including that of hunting-gathering peoples.

Critics view this attitude as a type of puritanism or ecological fundamentalism that sees humans as necessarily evil. They counter that there has never been a natural landscape in history that is pristine, one completely devoid of human influence (for good or bad). From this standpoint, forest people are a part of nature and must be considered a part of conservation efforts—as long as they can coexist with nature and live "sustainably."

However, this perspective has its own problems. First it creates its own philosophy of purity—the purity of human behavior, especially tribal peoples'. Critics charge this is a return to the "noble savage" ideal, which holds that "primitive" people live in greater harmony with nature and are thus morally superior to those of advanced industrial civilizations. The fact is that even these "noble" forest folk can wreak havoc on nature. And today, with smaller forests and larger human populations, this is increasingly the case.

A second set of problems is tied to the future of tribal people. If they do remain in the forests and continue with traditional ways of life, will this reduce their chances for better health, education, wealth, and other modern measures of development? Should they have access to anti-venom when bitten by a viper?

Also, by remaining in the forest and gaining protection there, will these traditional people simply be relegated to being artifacts—museum pieces conserved for the benefit of tourists and anthropologists? Finally, if the state allows tribal peoples to maintain their traditional ways of life, should this become a universal standard? What of all the deposed maharajas and the caste system, or all those dispossessed big game hunters who would still like a tiger skin on their wall? Then there is the traditional Chinese medical practitioner, who needs tiger bone to ply his trade. What of his rights? The list is long.

But the issue of people and wildlife in India will not go away. Both sides have valid arguments. It is a political as well as a scientific issue. How it will be resolved is still unclear.

◆ ◆ ◆

We stayed at Barheipani Waterfall for a day. It was picturesque setting, the water cascading down the sheer rocks and then disappearing into masses of giant boulders and thickets that obscured a stream that flowed at the base. The heat was stifling all that afternoon. Rabindra and I spent much of our time resting and taking frequent baths in the tiny cement bathroom of our lodge to cool down. We were anxious for the sun to set.

Once it did, we set out for the forest to search for animals. We soon encountered a small herd of sambar, large adults, grazing alongside the road. They bounded off soon after spotting us. The sambar is the most common deer at Simlipal because of the dense forest. And we soon saw more as we drove deeper into the woods.

Rabindra turned to me. "Maybe we will see elephants. They like these roads to walk on at night. But we must be careful. We do not want to scare them. Very dangerous!" he exclaimed as he raised his finger in the air. "Or maybe we will see the tiger. They like also like to walk on the roads at night."

I used my flashlight to scan the shrubs and trees alongside the road as Rabindra looked ahead. Periodically yellow and orange eyes of birds and squirrels lit up high in the canopy. Then there were big bright eyes closer to the ground—more sambar, which, like the proverbial deer in the headlights, stood mesmerized for a few seconds before disappearing in the underbrush.

The warm night and bright lights brought out insects—beetles and moths of every size, shape, and color bombarded us. At one point a giant water bug—almost five inches long—landed on Rabindra. He let out a brief shriek as I pulled it off. I shone my light on its long proboscis and told him it could inflict quite a bite. He was not interested and raised his arm in disgust as he turned his head away. I flung the bug high in the hair and watched its heavy armored body open up to reveal large delicate wings that sent it into flight.

Then we came upon a huge dark shape in the middle of the road. "Maybe an elephant," I thought out loud. Rabindra stopped and I looked closer. He got into reverse, ready to move quickly if need be. We both let out a sigh of dismay and relief as we saw it was a domestic buffalo. "It's from the tribal village over the hill," Rabindra said, shaking his head. "They let them out and they come right into the forest."

It slowly moved out of the way as we drove by, never looking up. We drove another half hour and then turned back, retracing our path.

Soon I saw something I had not seen before. At first I thought it was a pack of village dogs (curs), or maybe jackals. But they were larger, stronger, and more alert. The jeep stopped, with headlights beaming straight upon them. "Wild dogs," Rabindra whispered.

They were not afraid.—almost defiant—as they gazed our way and then slowly turned and trotted into the woods. We pulled up to the place of entry, and I shined my flashlight inside the open forest and spotted four of them drifting off into the distance, reddish brown creatures with a white underside and wide pointy muzzles, about the size of an American coyote, but far more streamlined—built for the chase.

The Indian wild dog or dhole (*Cuon alpinus*) is one of a number of canids found on the subcontinent, along with the jackal and wolf. But the dhole is unique, as it is a quintessential forest creature, just like the tiger. It is also a ruthless and efficient killer, hunting in packs sometimes 30 strong. Dhole run down their prey, coursing relentlessly, until the animal yields to exhaustion. Part of their strategy is constantly trading places. Those in the front retreat and rest, while those in the rear surge forward to replace their exhausted comrades. Almost no animal—not even the strong and agile sambar—can escape this constant assault. Once cornered, the animal is quickly killed and eaten. And a large carcass can be finished off in an hour.

Dhole usually avoid killing livestock, but they are still considered vermin by many Indian herders and farmers—perhaps because of their affinity with wolves and jackals. Under British rule they were ruthlessly exterminated because they killed so many game animals. Today the dhole is even more endangered than the tiger, as only 2,500 exist in all of Asia, the majority in India.

Unlike wolves, dhole almost never attack people. But there are cases where young elephants and grown tigers have been caught and overwhelmed by these fast-moving killers. I read of one account that took place in central India in the 1940s. It described the futile struggle of a single tiger against these unrelenting attackers. It was told to an Englishman by a shikari and his brother who witnessed the attack. The cat had already been fighting off the dogs for an hour:

> The tiger was in very bad shape and it was with great effort that he kept himself erect. Presently his head began to droop and again the dogs attacked, one fastened onto the tiger's throat and although immediately beaten to a pulp, its jaws remained locked and its hold could not be broken. In a supreme effort the tiger reared up on its hind legs with the dead dog still at its throat and others draped all over. . . . He toppled over backwards and was immediately covered by dogs, there were more convulsive struggles and all was still.
>
> By the time we got [back from the village] all that was left of the tiger was bones, sinews, and a few tufts of blood stained fur. We counted twelve dead dogs and could see where others had dragged themselves away.[5]

For the most part dhole and tiger respect and avoid each other. The dogs are generally diurnal hunters and the cats nocturnal. But in hot, dry weather like this, dhole will move and hunt after dark. What unfortunate creature would fall victim to them tonight?

We left Barheipani for Jorunda Gorge, where we arrived late the next afternoon. We would spend a few days at this place, the scenic centerpiece of Simlipal. It is a splendid example of raw tectonic power—an entire mountain torn in two, revealing the passage of eons in its many-hued walls. It was graced by a waterfall that cascaded 1,000 feet into a canyon, where forests had colonized pockets of alluvial soils deposited by a fast-moving river that continues to wear away the rock.

I wandered along the edge of the gorge, where a small weathered warning sign and a limp fence were the only barriers to a headlong fall into the precipice. I saw movement just below the rim of the cliff—thousands upon thousands of tiny, seething bodies. They were bees that had built their hives in these protective rocks. Few predators could reach them, allowing the insects to flourish. Their nests covered every square foot as far as I could see.

Rabindra soon joined me, and I pointed to the hives. "There are many bees in this place," he exclaimed. "There are so many in this forest. Now in this dry weather they are filled with honey. Sometimes they get this big," he said as he stretched his arms three feet apart. "High in the trees and under cliffs, that is where they grow big."

Then he went on to explain how tribal people are the expert honey collectors in these forests. "They are not afraid of the bees. They will climb right up to the nest as high as that tree," pointing to a hundred-foot sal behind us. "And they can get the honey without getting stung. Sometimes they use smoke and fire to keep the bees away. But they always leave a small part of the nest so that the bees do not leave. So every year they can come back and get more honey. All throughout this forest the tribes have their special honey places."

Long ago, before the spread of sugar cultivation, honey was an important commodity as it was the main source of sugar. It was a luxury for most people, except the forest tribes who always had ready access. Honey collecting remains an important part of these forest economies, and fortunately, it is a sustainable activity.

But wild bees can be dangerous. Forest travelers and even experienced honey collectors have been attacked and stung to death. The Indian giant bee, or rock bee (*Apis dorsata*)—the one that builds the large hives Rabindra spoke of—is notorious for its unprovoked attacks on people and animals. In some cases wild bees produce toxic honey from nectar collected from poisonous flowers.

Rabindra did not know whether the bees here in the canyon produced good honey or if they were aggressive. They were smaller than a honeybee, black in color, and agile flyers. We soon walked back from the edge of the cliff as the bees were getting more active and flying closer to us as the late afternoon heat intensified. More began to take flight, and soon there were hundreds of thousands, in frenzied flight in the air just above the canyon wall. We watched them from the porch of our lodge with awe and trepidation.

The bees finally slowed down as the sun set. I walked back up to the canyon at dusk and pulled out my binoculars to look at the hives. They were covered by thousands of tiny bodies all packed together, hardly stirring as night set in.

◆ ◆ ◆

The canyon was filled with mist every morning, which burned off completely by noon. I would wake up before daybreak to take in the majestic view and cool morning air before the heat set in. Animals were active this time of day, especially birds and monkeys. Birds chirped and called in chorus, culminating in the piercing "may-haw" of the peacocks. But all retreated back into the woods when the heat arrived, turning over the afternoons to the frenzied behavior of the bees and brain-fever birds.

One morning I spotted a small cat, slightly larger than a domestic feline. It was the secretive Indian jungle cat (*Felis chaus*). I almost missed it, as its dark brown coat blended perfectly into the rocky background. Only when it shot into the woods did I catch a brief glimpse. I noticed the Simlipal monkeys were also particularly wary. Perhaps they were unfamiliar with people, as there are so few visitors here. More likely they were too familiar with people—as hunters.

I asked Rabindra about hunting at Simlipal as we were out driving one morning. "Ahkand shikar!" he proclaimed in a loud voice. "Yes, this is the season for the great tribal hunts. They burn these forests and kill the animals. All tribes do it. It is their custom."

Then he grimaced, "They eat everything in the forest—rats, birds, snakes, monkeys. Big animals too!"

I asked him if he had ever seen the hunt take place. He shook his head. "They do not want outsiders to come. It is their ritual. The government tries to stop them. But they go far in the forest. The forest guards do not bother them there. Besides they are scared to be attacked."

We then drove on to a place where Rabindra told me the tribals had hunted a few years ago, before the government stopped them. We parked at a ridge and looked out across a neighboring hillside with open areas next to a thicket of trees and shrubs. "See there. That is the place the hunters stood and then they burned the woods over there," he said, pointing to the trees. "That whole piece was burned."

I could not see any evidence of it except the new growth of saplings and low shrubs. The forest can regenerate quickly. But it takes time and space, both of which are becoming scarcer every year.

One could clearly see the similarities between these more ancient tribal hunts and the later royal tiger hunts. Both used the principle of encirclement to drive the game into the open. One used fire and smoke, the other, the sounds of drums and horns to drive their quarry to their deaths. Neither gave the animals much chance of escape.

On our way back to camp, we stopped at another tribal hamlet to buy fruit. It was a familiar setting, with a half dozen huts tucked into a shady grove surrounded by sparse fields. Women and children were active doing chores. A few older men sat outside doorways. A group of younger men sat idle under a nearby tree, shy and interested in our presence, but a bit suspicious. Maybe they thought we were sent by the government to see what they were doing.

The scene reminded me a lot of Native American reservations in the United States, where young men—cut off from their hunting and martial traditions, which required physical endurance, bravery, and skill—now languished. Many turned to drinking and drugs (not uncommon here as well) or criminal activities as the only outlets to channel their libidos. Is this necessarily the fate of all premodern peoples?

What was the future of these people, I thought as we left. I looked back and they gave a slight wave, as did I.

And what of the tiger, once their respected forest neighbor—what is its fate in the changing world of the tribal communities and their ancestral forests?

The tribal question remains at the heart of conservation in India, especially the idea of sustainability. The question is not so much whether tribal people can maintain their traditional lifestyle in its entirety—something improbable and impractical. Rather it is a question of which tribal traditions can be maintained (or rehabilitated) in ways to benefit tigers and the forests that both share.

The essence of tribal life is its simple material existence, which may provide an alternative, even a foundation, for future conservation paradigms, because it stands in such stark contrast to the profligate materialism that is now ascendant the world over. By upholding this alternative, the tribal peoples are living witnesses to the fact that not all must be seduced by the siren song of material progress and the conquest and plunder of nature that it requires. From this standpoint, these "primitive" people may be more forward looking than many of the world's more "advanced' societies.

— 8 —

Man-Eaters of the Mangroves

Under normal circumstances, people and tigers can coexist. If people leave tigers alone, tigers will do the same. But there are exceptions. There are man-eaters. And one of the oldest and most fascinating questions about tigers is why they do eat people.

The most basic answer is injury and old age. Cats unable to hunt agile forest animals turn to easy prey like people and livestock. Another theory is that tigers are used to being around people. Familiarity reduces fear, which leads to curiosity and, when mixed with hunger, leads to man-eating. Places where tigers and people live in proximity, like India, or China of the past, have more man-eaters than places where they do not. The African lion, which generally lives apart from people, is less likely to become a man-eater than an Indian tiger. This is also the case with the African leopard versus its Indian counterpart.

These theories adequately explain most cases of man-eating, except one—the man-eating tigers of the Sunderbans, the mangrove forests on the India-Bangladesh coast. Dozens of people are killed there every year, more than any place on earth. These aggressive cats hunt people in forest and village alike. They have been known to drop out of overhanging trees on unsuspecting travelers and swim out to anchored vessels at night to set upon the sleeping crew. They will attack at night or in broad daylight. Nothing seems to stop them.

These animals are not old or injured, and they live in the least populated forest in India. So why do they kill people? Some believe it is a lack of suitable prey species. Boar and spotted deer are the only large animals in the Sunderbans, but they are not common. Thus, hungry tigers that live on a paltry Sunderbans diet of fish, crabs, birds, monkeys, and snakes will eagerly eat the only other large creature around—people.

Some speculate that the powerful typhoons that have assaulted this coast for millennia and left behind hundreds, even thousands, of human corpses at a time have given the cat a taste for human flesh. Others yet say that this harsh marine environment, and the amount of salt these tigers consume, has made them just plain mean.

Whatever the reason, man-eaters have lived in the Sunderbans forests as long as anyone can remember. Man-eating has been part of local myth and tradition for centuries. It was one of the first things observed by European explorers. French explorer François Bernier toured these forests in the mid-1600s and spoke of them.[1] During British rule, cases of man-eating in the Sunderbans frequently made headlines in nearby Calcutta. One famous case was the untimely death of a young man in 1787:

> A large fire was lighted, and an agreement made that two of the number should keep watch by turns, to alarm the rest in case of danger, which they had reason to apprehend from the wild appearance of the place. It happened to fall to the lot of one Dawson, late a silversmith and engraver in this town [Calcutta], to be one of the watch. In the night a tiger darted over the fire upon this unfortunate man ... In the morning the thigh bones and legs of the unfortunate victim were found at some distance; the former stripped of its flesh and the latter shockingly mangled.[2]

Two hundred years later, the former director of the Sunderbans Tiger Reserve, Dr. Kalyan Chakrabarti, gave a similar account:

> Sanyasi Mandal (a fisherman) was saying: "I opened my eyes from deep slumber. It was about 11 P.M. A pair of eyes was glowing in the poor, flickering light of the kerosene lamp in the boat and a warm putrid breath could be felt. I tried to shout but my voice failed. I have seen many tigers but nothing like the one I saw this time."
>
> It was customary for the fishing folk to get under the bamboo of the boat after the day's work. On this day too, they had sat there and chatted for hours before going in. Sanyasi cannot recall how he suddenly woke up to find the man-eater towering above him and his teammates right under the canopy. The animal selected Ananta Mandal, a man of 40, lifted him up in his jaws and jumping out of the boat, vanished in the darkness. Nothing could be done, as it all seemed to be over in a second.[3]

Despite their ferocity, Sunderbans tigers were never hunted down, and there was never an attempt to exterminate them because it was simply impossible. The Sunderbans is so vast and impenetrable, so totally hostile to human beings, that tigers cannot even be tracked, let alone killed. Maybe this is another reason they are so fearless. They know they can never be caught.

The Sunderbans is the last place on earth where the tiger alone is the undisputed king of the jungle. Not even mischievous man can challenge him there. But there is more to the Sunderbans than deadly monsoons and man-eating tigers. It is one of the earth's great natural marvels, as beautiful as it is frightening.

◆ ◆ ◆

The Sunderbans is India's last untamed wilderness—thousands of islands covered by mangrove forests, stretching over 100 miles along the India-Bangladesh coastline. The islands are small, most under a square mile, separated by a network of channels, lagoons, and estuaries. Each island is a mass of vegetation that grows so densely that little open ground remains. In some places scientists have counted up to 5,000 stems per acre. It is truly a jungle and a perfect habitat for the tiger, allowing it to sleep, prowl, and attack virtually undetected.

The Sunderbans owes its existence to the Ganges River, which deposits millions of tons of silt and debris into the Bay of Bengal every year. These sediments accumulate to form islands that are quickly colonized by seeds and shoots that then grow into forests.

The large amount of sediment and debris carried by freshwater rivers into the warm ocean creates a life-sustaining nutrient "soup" that supports a tremendous variety of aquatic life. The interaction between salt and fresh water is another dynamic aspect of this ecosystem, as these brackish waters are able to support both marine and freshwater species. The shallow waters team with crustaceans and mollusks, and the deep channels are home to many types of fish.

Like all coastal environments, the Sunderbans is in constant flux. Fierce storms and frequent tidal waves destroy and rebuild islands regularly. Mangroves quickly colonize islands, stabilizing their soil. But even these protective forests eventually succumb to the power of the ocean.

This instability is essential to the functioning of this rich ecosystem and is also what has helped protect it. Shifting islands, dense forests, and a lack of fresh water and arable soil have made most of the Sunderbans islands uninhabitable for people. The only exceptions are the larger islands in the mouth of the Ganges, which, together with the adjacent coast, are home to hundreds of farming and fishing communities. These people enter deep into the mangroves every day to fish the waters and collect wood, herbs, and honey from the forests. And every day they trust fate they do not fall victim to the most dangerous tigers in the world.

◆ ◆ ◆

It was now late spring when I left Kolkata for the Sunderbans. The intensifying heat already gave me a hint of the coming discomforts of the dry season—the fetid marine air, heavy with odors of salt, mud, and rotting things and active with insects. Then there is the tropical sun, its unrelenting glare so different from the temperate sun, which is welcomed and celebrated. The only reliefs from it are the clouds that build off the warm ocean in the afternoons, the shade of trees, or sleep in some dark, steamy hut.

The Sunderbans Tiger Reserve—the main protected area in the Sunderbans—lies just at the edge of the last inhabited islands and encompasses 1,000 square miles of mangrove forest. To get there from the mainland is not easy. I had to cross over a succession of islands by canoe, motorboat, and ferry. Each island had a bustling little port—a welter of colors, sounds, and smells—a place of commerce and a landing to receive the constant flow of maritime traffic. Sometimes I hired rickshaws or cattle-drawn carts to take me across the larger islands to reach the connecting port on the opposite side.

Some of these islands were a few miles across, revealing flat, rich soils planted with rice, vegetables, coconuts, and papaya. Villages were spaced intermittently along the intricate network of paths and narrow paved roads that covered each island. The houses were all built of dark gray, almost black, mud, of which there is no shortage here. It bakes hard in the dry season and can stand up to heat and storm. Each hut is covered by a neatly cut thatched roof and usually surrounded by groves of shade trees, including many fruit trees.

Although crowded, these islands are some of the most tranquil places in India, because there are no cars. And there are no cars because there are no connecting bridges. Island traffic consists entirely of rickshaws and bicycles, with the occasional motorcycle or tri-wheeler—something small enough to fit on a boat. Most people simply walk. And all islands are covered by narrow roads and footpaths that are artfully laid out to accommodate this more relaxed mode of transport.

The survival of this agrarian system is completely dependent on a vast network of canals, ponds, and dikes. Canals and ponds provide fresh water. Huge earthen dikes surround each island, keeping rising ocean tides, and especially storm surges, at bay. But despite all efforts to contain the constant threat of salt water, it is never wholly successful. Dams are often threatened by harsh weather, and even the drinking water here is never completely free of salt.

During low tide, the outer walls of the dikes stand 15 feet above sea level and are pure mud. The negotiation of the greasy gray matter requires not only steady feet, but also a resolute indifference toward mud-drenched shoes and clothing.

Historically, these large islands and the adjoining coast were covered in lush forest and grasslands that were home to many large mammals, including elephants, swamp deer, buffalo, rhinoceros, and tigers. Even the Javan rhinoceros (*Rhinoceros sondaicus*), now extinct from India and confined to a few forested tracts in far-off Indonesia and Vietnam, was once common here. But these coastal wildlife habitats and their wild inhabitants have slowly been lost to agriculture over the last century. The Sunderbans, like the duars to the north, represents the last fringes of West Bengal's once-extensive wilderness.

I had taken a large motorboat that held about 20 people on the last few miles of my approach to the Sunderbans Tiger Reserve. Most passengers were school children who are ferried through these channels every day. And every day they muddied their neat blue and white uniforms as they plodded through the thick ooze to and from the waiting boats.

We spent an hour moving through narrow channels, constantly stopping and dropping off passengers. After the last had disembarked, the boat hit full throttle as we entered an open bay. As the vessel approached the brackish waters, the wildness of the Sunderbans became immediately apparent. Each island was a mass of vegetation with no hint of human interference. To the casual observer, each is a uniform tangle of green. But botanists have classified a dozen distinct forest types here based on differences in moisture, salinity, and elevation.

We passed some large islands with some elevation; one had a rare grassy clearing that went all the way to the shore, where several spotted deer grazed and a solitary pig wallowed and rooted in the rich ooze revealed by the low tides.

The boat slowed as we rounded a large island and soon a wide wooden dock came into view, behind which stood the lone tourist lodge in the Sunderbans, a place called Sajnakhali. It was a weathered wooden building, built on 15-foot stilts to protect it from an unpredictable ocean. A 20-foot steel link fence surrounded the entire place—to keep out tigers.

Soon after I unloaded at the dock, I went to the front office, where I met an older man who ran the front desk. I asked him about tigers, especially the fence. He told me, in broken English and with hand signals, how a tiger had come just a few nights earlier and desperately tried to climb up the fence. It clawed its way halfway up and hung there snarling before a staff member threw rocks at it and frightened it away. It came back an hour later and tried again. The old man told me, with a reassuring nod, that tigers could not climb higher than halfway. Fortunately, the engineers had measured correctly.

I was given a small room that opened out to a wooden boardwalk that surrounded the lodge. The accommodations were rudimentary, and the meals in the single canteen were equally so, usually consisting of rice and dal (lentils) followed by fruit. The water and tea were always laced with salt, as truly fresh water could not be had here in the middle of the mangroves. It took a while getting used to and made one keenly aware how harsh this forest is for its inhabitants—both people and animals.

I stayed up late that night, kept awake by the barraging mosquitoes and also the thought of tigers on the prowl. More than once I walked outside to take a look at the steel fence, gleaming in the moonlight. It was all that separated me from the black mangroves beyond and the Sunderbans tigers.

◆ ◆ ◆

The only way to see the Sunderbans is by boat and hired guide. The guides are knowledgeable, as all live in the surrounding villages, and the boats are also locally owned, usually fishing vessels. For a reasonable fee they will take you out, alone, for an entire day.

It was already hot when my boat arrived that next morning. It was an old wooden structure that belched heavy diesel fumes from an all-too-short smokestack, and the fumes kept filling the small cabin where I sought refuge from the heat and sun. The captain was a fisherman, in his late sixties but wiry and alert. He was accompanied by a younger man, in his thirties, who worked as a schoolteacher and gave tours in his spare time. He had been working with the project for a few years, but as a lifelong Sunderbans resident knew its islands and backwaters intimately. His name was Kumar.

We stopped to pick up some supplies—water, vegetables, and rice for lunch—at a nearby dock and then headed off through a wide channel with olive green islands on every side. We soon passed an enclosed estuary where a half a dozen fishermen were busy at work. They labored in small crews aboard sturdy handcrafted wooden dinghies that were powered by paddles and long, oar-like rudders. Most used nets, but drop lines and traps are also employed, especially to capture crab and shrimp. Kumar explained some local fishing folk use even less conventional means—trained otters.

Apparently it is an age-old tradition. The adult animals are kept on long tethers tied around their waists and then released into the water to follow their natural instincts. Joined by their free-swimming offspring, the animals move in tandem and herd schools of fish towards waiting nets. After their chores are done, the animals get a small share of the catch. Some people find this ancient custom to be cruel. But the animals are generally well cared for, and it is a perfectly good use of local resources.

Whatever its form, fishing is central to this subsistence economy. And fishermen know the Sunderbans better than anyone, as they ply its network of waterways every day and through every season, to pursue their livelihood. They know its secrets and its beauty and also its dangers. Scores die here every year.

Many are killed by tigers when they enter small coves and creeks, which are good fishing spots but dangerously close to where tigers prowl. Hungry tigers will attack boats as they ply these narrow channels. The indefatigable cats will swim out into the middle of wide channels to board boats—even in broad daylight. So every fishing expedition is dangerous. Fishermen also get caught in storms, which sometimes come out of nowhere and can easily submerge their tiny fishing craft.

In the past, before advanced weather warning systems were developed, fishermen and entire fishing communities were engulfed by monsoons. Severe storms create enormous storm surges that reach miles inland because of the low elevations, inundating even farming communities. In some cases the death tolls were in the hundreds of thousands.

There are also poisonous snakes here. Underwater, there are sea snakes that are frequently caught in fishing nets. And on land, serpents are thick in the trees and the dense underbrush.

Sharks are also common. Kumar pointed back towards the fishing boats. "Sharks will attack these small boats, even here in these shallow channels." We both looked down at the murky water, wondering what danger lurked beneath. "It is especially bad when boats capsize or are caught in storms," Kumar explained. The biggest and most dangerous sharks here are (naturally) the tiger sharks. They are 20 feet long and known man-eaters.

"There are even pirates here," Kumar went on, as his eyes opened wide. "Especially on the Bangladesh side. They use fast motorboats and hide in the coves. They will steal anything, even fish from fishermen. Sometimes tourists are attacked. It is impossible to catch them. They know the Sunderbans well. They come and go with great silence and speed. Just like a tiger," he said with a laugh.

The boat slowed as we turned down a narrow channel about 30 feet wide and straddled on either side by dense growths of mangrove trees.

The Sunderbans forests are called mangrove forests because the mangrove tree (genus *Rhizophora*) is the dominant species. But the mangroves only grow along the shorelines, where they are uniquely adapted to the constant flooding and erosion. These are poor soils, highly acidic and low in oxygen. As a result, mangrove tree roots do not penetrate deeply; rather, they spread horizontally, creating a wide platform anchored by many smaller shallow roots. This stable foundation allows the tree to withstand wind and floods and grow to maturity. They also stabilize the islands by slowing erosion.

Mangrove trees also have intricate lateral roots called pneumatophores that take oxygen directly out of the air. On some species they are hairy masses that grow on the upper roots and branches. Here they are short spikes that grow up from the main roots lying just below the surface. At low tide, they become exposed and jut up out of the muddy flats in the hundreds, forming a formidable protective barrier around the tree.

Another key to mangrove success is reproduction. They develop dense elongated seeds, several inches long, that sprout leaves and roots while still attached to the tree (a process known as viviparous reproduction). When they reach maturity, they fall to the ground like a knife, embed themselves in the mud, and then continue to grow. Seeds that fail to implant float out into the open ocean, where their protective casing allows them to remain at sea for over a year. Some eventually take root hundreds of miles from their original destination.

Mangrove trees also store water in their thick, waxy leaves, which contain glands to secrete excess salt. As such, they can remain submerged in salt water for long periods—a perfect adaptation to constantly changing tides and frequent storms. The leathery leaves are also unpalatable to most animals but grow in dense clusters that provide shelter for them.

We passed the mangroves and then slowed to look up a narrow creek that cut deep into the island and then disappeared in a tangle of vegetation.

Kumar waxed poetic as he pointed to the masses of short palms that lined the narrow creek. "This is Phoenix palm, a famous tree here. We call it 'tiger grass' because tigers love them. They give shelter and create a network of underlying tunnels where tigers can move easily. It is a perfect place to hide and attack a pig or passing boat. Tigers like islands with lots of tiger grass. Like this one." I glanced nervously as our vessel glided only a few feet from the wall of palms, its welter of narrow leaves refracting sun and shadow, creating the perfect camouflage for a striped cat.

Sunderbans tigers are well adapted to this harsh climate. Despite their aggressive disposition, these are the smallest and leanest tigers in India. As such, they require less food and are more suited to these dense forests. They also have thinner coats to deal with the constant heat here and are more faded in color.

Sunderbans tigers also have enlarged kidneys and livers, perhaps because of the large amounts of salt they ingest in their food and water. And these tigers must consume an awful lot of salt, as there is hardly any fresh drinking water available, especially in these outer islands. The situation only worsens in the long dry season.

They also have slightly splayed feet with broad footpads that allow the animals to walk through oozing mud flats and loose sand banks, as well as swim—all perfect adaptations to the mangrove forests. This divergence in size and shape has led some tiger experts to conclude these cats may become a separate race in time, especially if they remain genetically isolated from other tigers.

Tigers are the only large land predator in the Sunderbans. Interestingly, the leopard, which shares every other habitat across the subcontinent with the tiger and still survives in places tigers no longer exist, is absent here. Whether the more aggressive tiger simply drove the leopard out or was hardier and more adaptable is unclear.

The boat then entered an open lagoon where the captain slowly accelerated. We passed some larger islands with high ground, where groupings of taller golpata and date palms grew in tandem with other tall trees that had long, graceful branches adorned with silvery leaves.

"You see that tree," Kumar yelled over the din of the motor, now at full throttle. "That is our most famous tree—the sundri tree. The Sunderbans is named after it. 'Sundri' means 'beautiful' in Bengali." And the sundri is really the only tree that garners that name in these otherwise stunted marine forests.

The sundri (*Heritiera fomes*) is, relatively speaking, a handsome tree. Its height—it grows to 70 feet—and long, splaying branches covered by longish glossy leaves make it stand out amongst its dull and stunted neighbors. It also has many uses. The tree's hard, reddish wood is highly desired for boat and furniture making. Because of this demand, the sundri has become less common on islands closer to shore.

It was midafternoon and time to eat lunch. Kumar directed the captain to take us to a shallow lagoon, where we came upon several fishing boats. We approached one, and Kumar began to negotiate with the fishermen for some of their catch. For a few rupees we got four small fish and two crabs, which Kumar proceeded to fry up in the cabin on a small cooking stove, after we anchored in a shallow spot near a thicket of mangroves.

As we ate our fish and rice, I noticed the tide slowly rising. But the mud flats below the mats of spiky mangrove roots were still exposed, revealing hundreds of tiny mudskippers. These unique fish can breathe air through their moist skins, which allows them to forage on land. But this also exposes them to predators, especially birds, whose periodic aerial attacks whipped them into a frenzy—causing hundreds to hop frantically towards the safety of the water.

After lunch we plied through more channels and estuaries. Rhesus macaques were common, moving deftly across the matted mangroves where they searched for fruits and seeds. They were equally at home on the shore, scrounging for washed-up food items or wading into the shallow water to hunt for mollusks and crustaceans.

I then caught sight of a large black and yellow lizard, slithering slowly over another clump of mangroves. It was a four-foot water monitor (*Varanus salvator*). It is a perfect predator. It is streamlined with a narrow, pointed head equipped with strong jaws. The short, muscular legs and long and powerful tail allow it to move quickly on land and swim with ease through the strong ocean currents.

The captain cut the engine and we floated silently to get a closer look. It did not yet see us. We watched it move intently, head weaving back and forth, tongue flicking in and out, searching for birds, bird eggs, or anything else it could swallow. But as soon as it spotted us it plunged into the salt water and disappeared.

"Some fishermen call these big lizards crocodiles," Kumar snickered. "But the real Sunderbans crocodiles are very rare. I have never seen one. Probably the captain has." He went down into the cabin to ask him and then returned to the deck.

"The old man says that huge crocodiles—man-eaters—were very common here when he was a boy. They were feared almost as much as the tiger. But now they are gone. Killed by poachers."

Kumar was speaking of the saltwater or estuarian crocodile (*Crocodylus porosus*). It is one of the largest reptiles on earth, reaching almost 30 feet. As its name indicates, it is perfectly at home in salt water, and it has been seen swimming calmly and confidently miles from shore. Yet it prefers coastal areas with brackish water, where the intermingling on land and water creates a perfect crocodilian habitat. It spends its days basking on the mud flats or navigating the narrow creeks and estuaries on the lookout for fish or any other suitable prey.

The saltwater crocodile is becoming rare throughout its former range, especially South and Southeast Asia.[4] Besides being hunted, it has succumbed to a loss of coastal habitat. Although they are still seen in the Sunderbans, they are extremely wary, avoiding people as much as possible.

The animals not afraid of people and that still thrive here in abundance are birds. The Sunderbans makes up one of the world's supreme bird habitats, an important rookery and feeding ground for dozens of species. Gulls, terns, raptors, wading and diving birds of every type appeared around every bend. Often flocks numbering in the hundreds flew overhead or were perched among the mangrove branches, sometimes covering entire trees in a collage of black and white feathers.

Many sea birds have a symbiotic relationship with fishing people, eating catch and waste thrown overboard. Even songbirds from shore come out to the islands to take advantage of the ample supply of seeds, fruits, and nesting materials.

"We should head out to the watchtower. You can see animals more there, deer and monkeys, sometimes even a tiger." Kumar turned to the captain and gave him the instructions. We were soon out in an open channel, one frequented by fishing and commercial vessels, that took us straight to the tower. It was perched directly next to the dock, which was entirely enclosed in heavy steel fencing—tiger protection. We opened the heavy gate and climbed up the narrow stairs about 25 feet, where we had commanding views of the flat, watery landscape. Dozens of islands, separated by an undulating maze of waterways, stretched across the entire horizon. It was an impressive sight, and this was only a tiny section of this extensive forest.

Ours was a larger island, with forest that surrounded a wide, grassy clearing in front of us with a pool of water in the center. "It is captured rainwater that helps the wild animals," Kumar explained. "The government cleared the forest and dug the pond to help tigers and deer. There are others like it around the reserve."

Except for the occasional passing motorboat, it was perfectly tranquil. Not even the birds, patches of white on the distant trees, made any noise. They periodically fluttered to a new roost nearby and then were still. At one point a spotted deer emerged. It moved cautiously, nervously, to the water's edge to drink and quickly retreated back into the brush. Only the breeze, soft and sporadic, was active.

After some time clouds began to build on the horizon, miles off, and the winds began to pick up. The captain, still down in his boat, yelled up at Kumar and pointed at the sky. It was time to head back. It looked like an isolated storm, as I could see sun and blue sky to either side of the billowing gray cloud. But in the Sunderbans one never knows. We picked up the pace, as the clouds closed in, but we made it back to the lodge in good time before a steady rain erupted.

Kumar went with me up to the dining room, where we got a cup of tea and talked awhile longer, as we watched the captain and his old boat drifting off to his island home through silver sheets of rain.

◆ ◆ ◆

I met up with Kumar a few days later and we took a ferry to one of the nearby fishing villages.

"All these people can tell you stories about tigers and almost everyone knows someone who has been killed by one," he explained as we approached a bustling little island port. "Some villages are known as 'widow villages' because of the number of women there who have lost their husbands to tigers."

Kumar then told me that he knows people in his (extended) family who have been killed by tigers. "But there is no anger," he said with a shrug and half smile.

This is the case throughout the Sunderbans. Despite the death and suffering, the tiger is not hunted down, or poisoned, or even despised. Man-eating is part of fate. It is the will of the gods. Rather than seek revenge against the tiger, people here have learned that it is better to remain faithful and humble. Living in the Sunderbans means danger and death. These fatalistic attitudes have helped tigers and humans to coexist peacefully for so many generations.

We soon disembarked from our ferry on the crowded main dock. Kumar led the way as we hurried through the mass of people and headed towards a small pier, where a group of men were climbing into a long canoe covered by a canopy and loaded with baskets and boxes, axes, and other work paraphernalia. We walked out onto the dock and watched them as they pulled out into the bay. Their boat, powered by four men with long wooden oars, moved briskly and silently out into the open water.

"These men are going into the forest for wood, maybe some fruit and herbs, but especially honey. These forests have very good sweet honey. It is the best honey," he proclaimed with pride.

Kumar went on to tell me that the dry season is the honey season, when the forests are filled with combs hanging heavy with the dark, sweet liquid. Honey is a valuable commodity, and honey collectors reap the rewards for their hard work. But it is probably the most dangerous work in the Sunderbans.

The collectors will spend several days out in the mangroves. They sleep on the boats at night, which are anchored well away from land—to protect them from tigers. The men move to land in the morning and receive their blessings from the accompanying holy man—either a Hindu *gunin* or Muslim fakir, who perform the proper prayers and rituals to protect them from storms, evil spirits, and most of all, tigers. Once inside the woods, they must work quietly, quickly, and always remaining vigilant, for the stealthy tiger can move to within a few feet of an unsuspecting worker and snatch him off without a sound.

"It is very hazardous work because they must go right into the tiger's home," Kumar explained. "Many men are killed like this. The tigers are so clever. They follow a line of men and then attack the last man and drag him off. The forest is so thick the tiger is sometimes only a few feet away from the men. They can attack so quickly and silently that the others do not even notice what has happened. The tiger loves to attack from behind. The only way to protect yourself is to always have someone guarding the back of the line."

He explained that the government has come up with a unique plan to protect people from tigers. They have given the woodcutters masks to wear on the back of their heads to fool the tigers. The cat thinks somebody is watching him and so does not attack. It has worked. But recently the cats have become wise to the trick and have resumed their attacks.

"It is really a struggle," Kumar exclaimed. "You see these people must go into the forest for their survival. They need the forest. They must go there even if the tigers attack."

We walked up an incline away from the water towards the bustling village. I looked back at the woodcutters, but they were gone, having rounded the bend and now moving straight for the mangroves. We walked up past the stalls and commotion to a quiet grove of trees on the far side of the village square. There was a small shrine where a few older women were praying and burning incense. Kumar walked up, hands folded in prayer, and bowed. We stood and observed for a few moments, and then he whispered that we move back as he wanted to explain its significance.

"This shrine is to Daksin Ray," he exclaimed while pointing to a small figure of the god in the middle of the shrine. He was a bare-chested man adorned in colorful pants, holding a bow and arrow, and seated on a tiger, which is his vehicle. "He is said to own all of the wealth in the Sunderbans and controls its evil spirits and those of crocodiles and tigers. He can enter the body of a tiger at will," Kumar added.

Kumar went on to explain that Daksin Ray invokes both admiration and fear because he is a god of wealth and wrath. When the big cats attack people, it is believed those people have insulted Daksin Ray or have failed to properly prepare themselves before entering the forest, by neglecting the appropriate rituals and offerings. So these men who just left for the forests gave their offering here in the hope that Daksin Ray will leave them alone.

"Not all forest gods are bad," Kumar stressed. "You see there is also the goddess Bonobibi. Legend says she was born in the mangroves as a person but was abandoned by her mother and raised by a herd of deer. She remained protected in the jungle, was infused with its magic, and eventually grew into a goddess. Her brother, Sha Jungli, the jungle god, joined her. To this day this pair travels the Sunderbans to perform miracles and rescue people from suffering. Their spirit contradicts the wrathful Daksin Ray." But local lore says that the two spirits coexist peacefully and together reign over the Sunderbans.

Many Muslims live in the Sunderbans, especially on the Bangladesh side. I asked Kumar about their attitudes toward the forest and its tigers. He smiled and shook his head. "They also pray at these shrines! Often Hindu and Muslim will pray together. I think it is the only place in India where this happens."

Praying to idols and participating in pujas (ritual offerings) might seem a bit out of place for a Muslim. But the Sunderbans Muslims are Sufis, the mystical branch of Islam that believes in an immanent God who is present in human form, such as in saints and in nature. This is very different from the orthodox Islam found in northern India and the Middle East that stresses a transcendent God removed from human experience and nature. Sufism is much closer to Hinduism. This spiritual convergence is further cemented by the shared experience of danger that both peoples face in their daily lives. These factors have helped to maintain relatively peaceful relations between Hindus and Muslims in the Sunderbans, which is unique in a land where religious conflict has been the norm for generations.

There is also no other place in India where nature and culture are such obvious factors in protecting tigers as in the Sunderbans. These dangerous forests—with their tigers and storms, sharks, crocodiles, and poisonous snakes—have instilled great fear and respect in these people for the forces of nature. But their bounty and beauty have also instilled love and gratitude. As such the Sunderbans forests are the most unique of all of India's—even the world's—tiger forests. And it is here that they may survive better than any other place on earth.

The Sunderbans still sustains the largest contiguous tiger population on earth. The Indian population is believed to number at least 100 animals, with about half that on the Bangladesh side. But nobody really knows how many tigers there are here. Again it is impossible to count them, especially with the old pugmark method.

Conservation efforts by the government have helped conserve this formidable marine tiger and its habitat. The Sunderbans was one of the original Project Tiger reserves and was enlarged to become a national park in 1984. Since then, the government has proceeded with a slow and steady program of conservation and eco-development. Fishing and the extraction of forest produce is confined almost exclusively to the buffer zone of the 1,000 square mile reserve and is controlled through a licensing program.

The main conservation goal has simply been to maintain the marine habitat, which in turn would sustain the tiger. Despite the size of these forests, the growth of demand for its resources, especially wood, is slowly beginning to have a negative impact on them. Population pressures are growing. The mainland is home to hundreds of thousands of people, and the Sunderbans is less than 50 miles from an even more crowded and resource-starved Kolkata.

However, patrolling this vast area with a few government vessels is still daunting, and illegal timber harvesting is impossible to control completely. Around the reserve areas government surveillance has kept these activities somewhat in check. Other projects that are still in the long bureaucratic pipeline include beekeeping, village forestry, mushroom cultivation, aquaculture, the development of freshwater facilities, and alternative energy programs.

Shifting resource pressures away from the forest islands will ensure that they remain viable habitats. The greatest threats facing the Sunderbans are not just direct threats like timber felling or even poaching. There are also indirect threats.

In recent years shallow bays and estuaries have been converted into shrimp farms to feed export demand. Wealthy outsiders, who want to capitalize on the globalization of the Indian economy, have orchestrated much of this. And natural resource exploitation is one of the quickest and easiest ways to make money. Not only are native ocean habitats converted, but also the large concentrations of these farms adversely affect water quality.

A more troubling development is oil exploration, which is currently under way in the Bangladesh side of the Sunderbans. It is believed that there may be large amounts of oil beneath these mangrove islands. Full-scale exploitation would undoubtedly fill the barren coffers of the Bangladeshi government and the pockets of its leaders. But the impact on farmers and fishermen and the subtle balance they have created with this beautiful land would come undone.

Perhaps the greatest long-term threat could be rising oceans. If the earth's climate is indeed getting warmer due to global warming, as many scientists believe, the Sunderbans will be adversely affected. Some predict that half the Sunderbans islands could disappear. Others think it may be a slower process. But in either case, the tiger will certainly be imperiled. If this were to happen it would be a supreme tragedy. It would also be a terrible irony.

These inhospitable forests, which have endured virtually intact into the twenty-first century and have been able to protect tigers better and longer than any place on earth, would in the end still succumb to mischievous man.

—9—

Kipling Land

Central India has more tigers than any place in the country—over a quarter of the total population. Most are found in an area known as the Central Highlands, a series of thickly forested mountain ranges—the Vindhyas, Satpuras, Mahadeos, Maikals, and others—that stand fortress-like in the middle of the country, marking the boundary between the Gangetic Plain and the Deccan Plateau.

The Highlands range from 1,000 to 5,000 feet and represent a wide range of rock formations. There are massive volcanic extrusions with curious step-like shapes called "traps." Some mountains have been worn down by water and wind to form smooth hillocks and well-sculpted mesas. Then there are hills with fragile crag-like formations comprised of many-hued sedimentary rocks. And in other places, ancient lava beds have spread horizontally rather than vertically, forming vast sheets of rock barely habitable by plants or animals.

Central India's forests are also unique because they are where the country's teak-sal divide, which roughly follows the 80th longitude, is found. In general, sal grows east of this point and then spreads north and west throughout the Himalayas and southeast into the moister hill tracts of the state of Chhattisgarh, ending abruptly in Andhra Pradesh, where scrub forests are again dominant.

Teak grows west of the divide, stopping in the drylands of Rajasthan but spreading southwest throughout the peninsula, including the Western Ghats range. Except for a few isolated areas, the ranges of these two species are almost completely exclusive of one another.

Both climate and geography contribute to this dramatic ecological break. Eastern and northern India have alluvial soils, which are preferred by the sal. Teak thrives in the more porous and rockier soils of the south and center-west. Sal also prefers cooler climates, thus is found in the Himalayas, whereas teak is the more tropical species.

Water is another important feature of the Highlands, as it is the source of several of the country's largest rivers, including the Tapti, Mahanadi, and Narmada. Of these, the Narmada is the most significant. It is India's second holiest river after the Ganges, as it is said to have sprung from the body of the god Shiva, the destroyer.

Rugged terrain and distance from the populated plains and coasts have made the Highlands a haven for wild animals. Geography also gave central India a unique human history. The Highlands were home to many tribal peoples like the Gonds, Baiga, and Korkus. And for millennia, they were a place of refuge from tax collectors and tyrannical rulers. The Highlands were also the perfect hideaway for bandits, or dacoits, who preyed upon unsuspecting travelers. The most notorious of these were the thugs, who strangled their victims with saffron scarves, robbed them, and ceremoniously buried them, all as a testament to their faith in Kali, the goddess of death and disease.

Central India was also one of the last places to be explored by the British. It did not become well known to Europeans until the 1870s, after the explorer Captain J. Forsyth published *The Highlands of Central India*,[1] which described the natural history, geography, and human cultures in great detail. Forsyth died soon after its publication, the victim of illness and fatigue endured during his years in this remote land. His book remains a classic to this day.

After the British consolidated political power, central India remained semiautonomous, a concatenation of hundreds of princely states that upheld a feudal way of life until Indian independence. It was here that many Europeans came for sport or simply to enjoy the last vestiges of ancient India.

Artists and writers also came to behold this land of physical beauty where animals and people still shared a common forest existence. There was Verrier Elwin, who left the Anglican priesthood in the 1930s, moved to a Baiga village, married a tribal woman, had children, and wrote one of the more comprehensive books on tribal life before dying of liver failure due to excessive drinking. Central India was also the first place a comprehensive study of tiger behavior was conducted; the research was done by American biologist George Schaller in the 1960s. It remains a benchmark on the animal's life and habits.

But perhaps central India is best known as the place Rudyard Kipling chose as the setting for his *Jungle Book* stories. These stories introduced the secret world of the Indian forest—its animals and people—to the European imagination. Although children's tales, they contained a sympathetic and realistic perspective on a world little understood at that time.

These books were my first exposure to India and the tiger (even though Sher Khan was the evil one—no doubt a bias of the imperial British lion!). Nonetheless I read them over and over and was eager to see "Kipling Land" firsthand.

◆ ◆ ◆

There are two places in India that are believed to have inspired Kipling. One is the Kanha forests near the Seonee Hills, where Kipling's stories were actually set. The other is the Bandhavgarh forests farther north, but more accessible to the Gangetic Plain, where Kipling live and worked.[2] Bandhavgarh was my first experience in central India.

Bandhavgarh is in the northern rim of the Highlands in the Vindhya Mountains, an area known for its granite outcroppings and flat-topped mesas surrounded by mixed sal forests. It is a day's drive north of the city of Jabalpur, a pleasant city where I spent a few nights before taking a crowded little van to the park.

Bandhavgarh means "brother's fort" in Hindi, named after the fortress that was built atop the massive granite mesa that stands in the center of the park. Legend has it that the Hindu Lord Rama gave it to his brother, Laxma, to watch over the far-off island of Lanka (Sri Lanka). Built over 2,000 years ago, it was still occupied until the mid-twentieth century. Today a lone Hindu priest lives there, tending his temple, which is still visited by the local villagers.

Bandhavgarh was part of the ancient princely state of Rewa, one of the largest and most famous in central India. It was ruled by Rajputs who came from the west (today Rajasthan) in the twelfth century. Its mountain location and martial culture helped Rewa maintain semiautonomy and friendly relations with both Mogul and British empires. And it was one of the last princely states to become part of a unified India in 1947.

Rewa's long tradition of cultivating music and the arts also earned it respect. But it was most renowned for its abundant forests and many wild animals, especially tigers. The Rewa maharajas were some of the greatest tiger hunters, and the Bandhavgarh forests became a favorite hunting destination for sportsmen and dignitaries for centuries. It was set aside as a tiger sanctuary in 1968 and then expanded and made part of Project Tiger decades later.

The Rewa forests also became famous after the discovery of white tigers about 50 years ago. They were not albinos, but rather a distinct color phase. Always rare, they are confined almost exclusively to zoos, where they continue to be bred.

Today Bandhavgarh is famous because it is one of the best places to see tigers in India. When I checked into the government-run bungalow, the first thing the manager told me—almost as a promotional—was that I would definitely see a tiger tomorrow.

"You are guaranteed," he said with a smile, as he handed me my keys. I wondered if it was money-back guarantee, as I had heard this pitch many times before. But his guarantee seemed good, because that evening as I sat out on the porch of my tiny cement bungalow, I heard the deep guttural roar of the tiger off in the distance—*aeow . . . aeow . . . aeow*. It resounded throughout the valley and went on for 15 minutes. And then there was total silence. I felt fortunate.

◆ ◆ ◆

The next day I saw four tigers within an hour of my first park visit. A tigress and three cubs were feeding on something in deep grass close to a forested hillside. I was on elephant back, which allowed me to approach within a few yards of the cats. They gnawed away on the bloody body, which turned out to be a young spotted deer. The only way I could recognize it was by the uneaten head. The eyes were wide open, the last terror-filled seconds of its life embedded in a glassy gaze.

The deer was probably killed at dawn. The tigress then spent a half hour "preparing" the family meal by first removing the fur and thick skin before consuming the favored entrails. The cats then ate the flesh and smaller bones, leaving the head, heavy bones, and unpalatable stomach behind for scavengers.

We arrived just as the cats finished their meal. Bellies filled, they lay down a short distance from the carcass and meticulously groomed themselves. Afterwards, they retreated towards the hills, probably to sleep off their meal, leaving the fleshless heap behind. Vultures had already gathered in the nearby trees, waiting anxiously for the cats and our elephant to move safely away. The elephant lumbered back to my waiting jeep, where my guide was ready to find more cats.

We drove out of the park center to the periphery, where we came into solid forest, occasionally broken up by streams and meadows. Most were sal trees, but they were smaller than others I had seen in India, because of the drier conditions found here.

The road skirted the edge of the woods. We drove slowly, looking furtively beneath the open canopy for movement. Many trees had already shed their leaves in the pre-monsoon heat, creating a vast crackling carpet of yellow and brown. Even the lithesome tiger must move with special care through these noisy piles when stalking the herds of deer that congregate to feed upon them—leaves are an important food source until the coming of the rains and the regeneration of vegetation.

The only activity was by small birds and troupes of langurs. We eventually came to an opening with a water hole, now at low ebb late in the dry season. My driver knew exactly where he wanted to go—a small clump of shrubs at the shore's edge where we could hide and get a good view of our surroundings.

As soon as we stopped he pointed to the opposite shore, where a large skeleton lay. "Nilgai," he said authoritatively. "It was killed a few days ago, by a tiger." It was stripped of all flesh, and the blood and sinews had already dried hard onto the bone. A few vultures stood around it, one actively poking his naked head deep into the rib cage to pull out some last fetid scrap. They were large birds, black with white backs, hence their name white-backed or Bengal vulture (*Gyps bengalensis*). At one point the busy one pulled out something long and stringy that caught the eye of the others, leading to a brief struggle before the victor swallowed it whole. They were very efficient.

My guide tugged on my sleeve and pointed to a nearby bush, where I saw movement. A faint gray shape emerged and slowly drifted towards the skeleton.

It moved cautiously, making a wide circle around the birds, which kept a close eye on their competitor. It was a hyena. It approached the nilgai's bony head and began to gnaw and pull, but it did not get much as the birds flapped their broad wings and made threatening squawks, and that scared him off.

The hyena retreated into the bushes and disappeared.

"She will be back tonight," my guide assured me. "She has a den close by. I have seen her here before. When it is dark she will come and eat more bones. The vultures eat by day, and hyenas by evening or night or early morning like now."

The striped hyena (*Hyaena hyaena*) is smaller, and obviously less aggressive, than its African counterparts. Like all the hyenas, it is an odd creature—half dog and half cat (although more closely related to the cats). The whole appearance is unclean and unkempt. It has a dull gray coat with dull brown stripes. Long dorsal hairs run down its back, emanating from an oversized head. But it has powerful jaws that can crack the hardest bones—even those left behind by the tiger.

It is still common in India, found throughout forests and farmlands across the subcontinent. Like most scavengers it gets little praise. Yet, together with the vulture, it performs the critical chore of keeping the countryside free of rotting carcasses, including those of millions of livestock.

On the second day I saw two more tigers—both big males. The first was at the same meadow where I saw the female and cubs.

We spotted him sitting erect in the yellow-brown grass at the road's edge. He was perfectly hidden, except for the bright white ear spots, which gave him away. We stopped and watched as he strode out onto the road 20 yards in front of us. He heard us but never turned around; only the prick of an ear let us know he knew we were behind him. This cat was used to people and big enough to fear nothing. He kept strolling down the road, slowly, occasionally pausing to survey his kingdom.

He was well over 400 pounds, with a massive head and a mouth that could swallow a human head whole, and then some. Despite his size, he moved gracefully. Large pads cushioned his every step; his solid muscles rippled subtly while his tail waved casually to and fro. We watched him until he disappeared around the bend and then drove slowly up behind him. He came back into view and then looked back, his deep yellow eyes fixated on us. For a second I felt like some hapless prey animal caught in his hypnotic trance. Then he turned and went on to an opening in the tall grass and disappeared. The stalks above moved ever so slightly as he moved off in the distance.

"He is the dominant male at Bandhavgarh. No other male will challenge him," my guide whispered. "He has three tigresses here, one with cubs [the ones I saw the previous day]. But there are two other males here in the valley. They live on the other side of the mountain." That was our next destination.

We soon came to a hilly place where the road curved and dipped and turned to heavy gravel. The terrain was rocky, marked by numerous outcroppings and ravines. Thorny shrubs were now dominant. We stopped at a small stream that had been reduced to a series of small pools by the dry weather. It ran through a solid granite slab, a massive volcanic outpouring that covered a few hundred square feet. The opening was about 10 feet wide and equally deep, but with many wide pockets, hollowed out by the water, below its rim.

"Tigers like this place in the dry season," my guide told me. "There is water and shade here," he explained as we got out of the jeep and walked towards the edge of the ravine, cautiously peering down into it. We walked all around the perimeter to see all sides, in the hopes of a tiger. He then pointed to large pool. "Sometimes the tiger will sit in that pool all day when it is hot," he said knowingly. But that day there was no sign of him.

We drove back to the other side of the mesa to a little wooded grotto at its base. It is famous for the 30-foot statue of the god Vishnu, called Shesh Saiya. The body lies prostrate, surrounded by a pool of water, and is in good shape considering it is a thousand years old. It was hewn out of solid granite, arduous work obviously done with the same devotion and fervor that fills those who have come to pray here ever since it was made.

Tourists also come to worship with their cameras and video recorders. I left just in time to avoid the onslaught of three jeep loads of Indian tourists, who piled out of their overcrowded vehicles and descended upon the statue with a mix of ancient veneration and modern entertainment.

We moved on to more peaceful surroundings—a massive granite outcropping where a dozen small holes—almost perfect circles—had been carved out of the sheer cliff wall. My guide told me there are 39 caves here of varying sizes. They were used by holy men who came there to pray and fast, even die.

"In my grandfather's day they still came here. Now no one comes anymore." He added with a hint of remorse, "Only wild animals. They like them because they are cool and give good protection." Then he smiled mischievously, as if holding back some secret. He then pointed to the far cave. "Look there. Can you see him?" I gave a skeptical glance.

He prompted me again, so I pulled out my binoculars and looked more closely. Sure enough. There lay a big cat in the shadows, dead asleep on the cool, sandy floor. I handed the glasses to my guide to take a look. He nodded his head in satisfaction and told me this particular male often came here. "This is his territory. These caves are his," he chuckled.

We watched and waited for a half hour to see if he would awaken. But he did not. At one point my guide even began to yell and got out of the jeep and threw some stones (not advised). But nothing stirred the slumbering giant. But it was a perfect experience nonetheless—the sleeping cat, safe and unperturbed by man—surrounded by the vast sal forest, now serene and sparse, deep in the dry season.

Three grey hornbills soon landed on the upper branches of a nearby tree—one of the few evergreens, making it stand out among its leafless deciduous neighbors. They were soon busily and noisily eating the large fleshy flowers that grew in abundance.

"That is mowha," my guide told me. "Birds, monkeys, and deer eat these flowers. Even people like them. For tribal people they are very important. They eat the flowers and fruits. It is also used to make 'country liquor,'" he said in a disapproving tone. My guide was obviously a Hindu, and the tribal custom of liquor making (and drinking) has always been frowned upon by them. But for local tribes this liquor is an important part of their culture and is used in almost every ceremony and ritual. The trees are carefully tended, even in the midst of the forest.

◆ ◆ ◆

I returned to the Bandhavgarh forests many times over the course of the next few days. I never saw the third big male, but I encountered at least one of the other six cats almost every visit. I saw the hyena again and many deer and monkeys.

I visited the small museum near the park entrance where the Maharaja of Rewa's old hunting muskets, some 10 feet long and all intricately carved, hung on the wall—now long silent.

And in the evenings, after dinner, I sat out on my porch and waited for that big male tiger to roar, and almost every night he did. And I felt fortunate, perhaps as so many who had come to these forests decades and centuries ago also had. Kipling would have agreed.

◆ ◆ ◆

Kanha was my next destination. I returned to Jabalpur for a fortnight and then took a taxi for the day's trip to the park, which lies in the Maikal hill range toward the southern edge of the Highlands.

My driver understood not a word of English, and with my few words of Hindi, the journey was spent in almost total silence. But he was an amiable fellow, an older man, humble and courteous, no doubt the result of a lifetime of Hindu training. The drive was also quiet as there was hardly any traffic, due to the relatively low population in central India.

We reached the countryside a few minutes after an early departure. Villages and farms were spread farther apart here because of the hilly terrain and rocky soils. Most were concentrated in the valleys that appeared intermittently amongst the rolling landscape. Homes were made of simple mud and thatch, and most farmers still relied on handmade wooden plows, pulled behind emaciated cattle. A few wealthy farmers had buffaloes, but they were rare in these parts. Modern farm equipment was almost nonexistent.

Forests were abundant, and large swaths appeared on every horizon with smaller parcels growing between farms and villages. Water was also abundant. We crossed several streams and even large rivers, including the Narmada.

The mighty Narmada originates in the Maikal Hills not far from Kanha, beginning as a veritable torrent as it winds its way down the steep and narrow gorges of these upper elevations, gradually widening and slowing as it flows into the flatter terrain of the west. It eventually makes its way to the Arabian Sea in the state of Gujarat, 800 miles away.

There are Shiva shrines and temples located along the entire length of the river. And every year millions come to pray and worship along its shores. The most devoted will make the round-trip pilgrimage of 1,600 miles—which can take over a year—as expression of their faith and devotion.

Our taxi slowed as we crossed over the narrow steel bridge to accommodate several groups of people making their way down to the water. I signaled to the driver to pull over at a small bluff on the opposite shore that looked out over the brown waters, still flowing swiftly, as we were not far from their mountain source. There were a few bathers that morning, and an old man seated in front of a small shrine perched directly below us at the river's edge. A hint of incense filled the air, mixed with smoldering dung. The setting was serene, a timeless image of India.

But this may soon change, as the river is now the focus of a new type of faith and devotion. Indian officials, industrialists, and international aid agencies have faith that this mighty river can be harnessed to provide electricity, drinking water, and irrigation for west and central India. They are deeply devoted to this dream, and with the help of the Narmada Valley Development Plan, they hope to create over 3,000 small and large dams along the entire length of the river.

To date, many have been completed or are close to it, including the massive Sardar Sarovar "mega dam" in neighboring Gujarat and the equally enormous Narmada Sagar dam in the heart of the Highlands. These and the other dams have already flooded hundreds of square miles of farmland and forest, including important tiger habitat. Thousands of rural people have fled their flooded homes, and hundreds of thousands more will be resettled when the project is completed. The real beneficiaries are industrial agriculture, big business, and the cities that continue to grow and demand resources.

As we headed back to the car I turned around to look at the Narmada one last time, tracing its path as it disappeared on the horizon. This mighty river, so long a part of Indian myth and history and the object of intense devotion, will soon to be turned into a series of placid lakes. How ironic that the Narmada is associated with the god of destruction. Maybe someday Shiva will take revenge.

We kept moving eastward and soon encountered more water—the first signs of the monsoon, which had already made landfall along the eastern coast. Clouds built on the horizon, and we drove through localized downpours that cooled the air temporarily. But as soon as they broke, the sun shone through and the humidity grew. And by noon the stifling heat of the dry season had returned.

The rains picked up again as we drove into the village of Khatia, the gateway to Kanha. Fortunately it was easy to find lodging, as the tourist season was almost over and the park was about to close with the onset of the monsoon.

My driver drove up to a lodge, one he obviously knew well, as the manager came out to greet him and quickly had rooms ready for both of us. The lodge was relatively new—a bit gaudy, like most modern Indian construction. But it was tucked into a pleasant patch of woods where a nallah, now slowly filling with water, cut through it.

The hotel manager was a middle-aged man with a large belly that hung well over his belt. He was garrulous, a smooth talker, but inoffensive. He also was well acquainted with Westerners and was eager to impress me with his knowledge about American music and movies.

He was soon joined by his staff. And since I was one of only two guests, they waited on me at every turn. As we walked through the rain to my room, with two boys in tow, haggling over the privilege to carry my single bag, he told me that they might shut the park early if the rains continued.

"The rains make it impossible to drive on the dirt roads. They are all mud. Besides, you won't see any tigers. They all head for the hills," he said nonchalantly.

Then he caught himself, realizing that I would leave if I could not get into the park. Noticing my dismay, he gave a sheepish smile and reassured me. "Don't worry. The rains will end. It is still too early for the monsoon. You will see tigers. No problem! No problem! I know the PM here; he will take you out even in the rain. I promise," he said with another embarrassed smile.

But it kept raining. It rained all throughout dinner, and I watched the rains through the cheap windows in my room, which leaked, until I went to bed. It rained all night. It rained all next day. At one point I had to get out so took a walk through the neighboring forest, now completely still except for the soft patter of drops, which came down lightly, but steadily. I walked along the nallah, which had now turned into a small creek.

Later that afternoon I had to get out again, so I fetched my driver, who had spent much of his time sleeping, to take me to town to visit some village shops. Afterward we drove through the nearby forest and farms at the edge of the park. I had to see *something wild* before the weather forced me to leave.

We drove into a large stand of woods, now dark, wet, and cold, where again nothing stirred except the large leaves on a low-growing plant that undulated slowly with the falling drops. We crossed several larger streams, which had swelled to capacity and carried dangerous cargoes of stones, branches, and plant debris. Water ran in sheets across open granite escarpments and over grassy knolls. Forest trails became small streams. The monsoon had arrived. Or so I thought.

Sometime that night the rains stopped. The next morning the clouds remained, but the air was clear and I could see the wooded hills on the horizon, still shrouded in early morning mist, for the first time. Delicate shoots of grass had appeared like magic on the dusty broken soils all around the lodge. Soon the trees would leaf out in lime and emerald green, and the entire land would be transformed.

It was obvious even from these few days of rain why the monsoon is so deeply woven into Indian culture—it is the life force of the continent. It comes with regularity every June and gradually soaks the entire country for months, lifting it from a dry lethargic haze to something magically alive. This short prelude created relief and rejoicing everywhere—even among the animals. Village dogs frolicked in the puddles, and rejuvenated cows indulged themselves upon the bounty of fresh grass.

For the tiger and other forest creatures, the rains bring temporary havoc but also renewed freedom. Water and fodder become so scarce in the dry season that forest animals are forced to live close to the few permanent water sources, increasing the risk of disease, predation, and competition with livestock and their owners. But the rains change this. The availability of grass and water allows herbivores to spread out across the land. Tigers too spread out, making them harder to find.

At breakfast the manager came to me excitedly and told me he heard from the park staff that if it stayed dry, the park would open up later that afternoon. The clouds hung low all day, but the rains did not return. Kanha at last.

We entered the park by jeep at midafternoon. The dark clouds created a gloomy atmosphere that only intensified as we entered the already dark woods. Inside the canopy the air was heavy with the smells of moist earth and wood, and the dirt tracks were muddy and filled with puddles.

Kanha's forests are mainly sal, mixed with equally large biju and haldu trees. Small spindly species cover the rocky hillsides, and bamboos and creepers are concentrated in the ravines. But the feature that distinguishes Kanha from other Indian sanctuaries is its many grasslands—over a dozen different varieties, both tall and short grasses; some hundreds of acres in size, others only a few; some natural, others man-made. Altogether, they cover over 15 percent of the land area.

Most natural grasslands are tall grass species found atop the flat mesas of the higher elevations and in the low-lying valleys. These latter types are the *maidans* (meadows) that have made Kanha famous. They are particularly rich ecosystems, with many grass species, all nourished by the water that drains there. Some grasses reach eight feet around streams and ponds, but most are less than half that height.

The man-made grasslands are mainly short grass varieties that grow on former village sites and farmland abandoned in the 1970s to make way for Project Tiger.[3] They cover some very large areas but overall represent only a small portion of Kanha's grasslands. The government has worked hard to maintain and even enlarge all the park's grasslands to increase the herbivore population.

As a result Kanha has the largest and most diverse herbivore population in India—five deer species, three antelope, the gaur, and the ubiquitous wild boar.[4] As such, it has often been likened to a mini-African savannah. Ample prey, water, and cover are what make these forests some of the best tiger habitats on earth.

The tour of Kanha soon took on a distinct pattern. We would drive through dense forests that would suddenly open up to bright green meadows. We would then stop and watch for animals. Spotted deer were by far the most common animals at Kanha. Huge herds, often numbering in the hundreds, appeared in almost every maidan. Gaur were also common; the secretive sambar and barking deer were less so. Troupes of langurs coursed across the tops of the surrounding woods, ever vigilant for a tiger ready to burst forth and grab a victim that ventured too close to its forest hideaway. These "ecoclines," where wood and meadow meet, are where tigers are frequently seen at Kanha.

This was the heart of India—Kipling land. The landscape is not majestic, something that fills one with awe, like the Himalayas. Rather it is poetic, evocative of some hidden beauty. This was captured best by the maidans, each having an atmosphere of secrecy and surprise whenever encountered. This is the magic of these forests.

For European visitors of the past it was also a refuge from the crowded plains and the demands of empire, a place lost in time, where the sleepy rhythms of the tribal village or the opulent splendor of isolated kingdoms still held sway over the rule of money and machines.

It was also familiar. The orderly woods, the open meadows filled with deer and wild boar, and the fox-like jackal were reminiscent of Europe. It was a pastoral setting that, perhaps, fulfilled a romantic longing for home.

But this was not an English park. It was still a wild forest filled with cobras and vipers, bears, wolves, leopards and, of course, tigers—lots of tigers. And perhaps it was this paradox—the meeting of the pastoral and primitive—that made these forests so alluring, even today. There was also the knowledge that someday this magical forest, this refuge and retreat, would be conquered.

Rudyard Kipling became famous because he was able to render these ideas in his art. He was certainly an imperialist and defender of empire. He has been criticized as an "orientalist," one who perpetuates stereotypes like the "noble savage," represented in characters like Mowgli of *The Jungle Book*. But he also depicted this world with great accuracy, a world in which people still lived close to nature, a world with privileges and perils, as is depicted in this passage from *The Jungle Book*:

> Mowgli went off to a circle that met every evening on a masonry platform under a great fig-tree. It was the village club, and the head-man and the watchman and the barber (who knew all the gossip of the village), and the old Buldeo, the village hunter, who owned a Tower musket, met and smoked. The monkeys sat and talked in the upper branches, and there was a hole under the platform where a cobra lived, and he had his little platter of milk every night because he was sacred; and the old men sat around the tree and talked, and pulled at the big huqas (the water-pipes), till far into the night. They told wonderful tales of the gods and men and ghosts; and Buldeo told even more wonderful ones of the ways of beasts in the Jungle, till

> the eyes of the children sitting outside the circle bulged out of their heads. Most of the talks were about animals, for the Jungle was always at their door. The deer and the wild pig grubbed up their crops, and now and again the tiger carried off a man at twilight, within sight of the village gates.[5]

We drove on to another maidan where we watched a large herd of spotted deer, grazing peacefully as it slowly moved across the open until it became obscured by a tall stand of grass. Today they would not be carried off by the great predator. Today, at least, they were safe.

Soon we drove on to a vast clearing, where a village once stood. Faint hints of human habitation still existed—the outline of random paths, areas where crops grew, and raised places where houses stood—a place once busy with human activity, now quiet, receding into uniform stands of grass.

"This was a Gond village once," my guide burst out. "They roamed all over these forests." He pointed to the surrounding hills. "Even the roads we are using now—many were old hunting trails. The Gonds now live on the outskirts of the park where they still farm. But hunting is strictly forbidden."

The conversation then turned to animals. "This is where you will see antelopes. They like these open places with short grasses," he explained. We waited and watched and spotted a lone black figure grazing far in the distance. It was a nilgai (*Boselaphus tragocamelus*), a large nondescript cow-like antelope. The nilgai is also known as the blue bull, because its large size, smallish head, and tiny horns make it look more like a type of cow than an antelope. This is also why Hindus revere and worship it in many parts of India, thus ensuring the animal's survival.

I asked the guide about the blackbuck, another Indian antelope.

"No," he shook his head. "It is rare here. Some say it is now extinct at Kanha. I would like to see one. It is very beautiful. But I have only seen it in pictures."

India only has five antelope species, which is dwarfed by Africa's tremendous diversity. But it can be argued that the Indian blackbuck (*Antilope cervicapra*) is the most beautiful of all the antelopes. The males are the size of a gazelle, black with a white underside and long, spiraled horns. Blackbuck often travel in large herds, where dozens of pointy horns moving in unison further enhance their majestic presence.

Although still common in western India, the blackbuck is naturally rare in central India because of the limited habitat, as it prefers the dry open plains and the edges of farmland. Naturalists speculate the blackbuck came to Kanha by following human cultivation. And with the creation of the park, a small population was stranded and cut off from others. But it is declining (or may be gone) because short grass ecosystems are less common here. The government also promotes tall grass conservation and expansion, as the more abundant deer prefer them.

It was now late, so we headed back. As we sped along the open dirt trails to the main gate, the guide turned and assured me that tomorrow we would see more animals. "We will see the tiger and maybe, if we are lucky, the barasingha. Do you know of the barasingha?" he said with a twinkle in his eye. "It is very famous at Kanha."

I did know the barasingha but had not yet seen one. It is a deer known for its elegant antlers that grow in a semicircle with many forward-pointing tines. That is why it is known as "barasingha," which means "twelve-tined" in Hindi—although large males sometimes have over 20. It is handsome creature—chestnut brown with a cream underside and a dark stripe down its spine—and is fairly large, reaching 400 pounds.

It is also known as the swamp deer, as it is most widespread in the floodplains of northern and eastern India. Those found here are a separate subspecies, the "hardground" barasingha (*R. d. branderi*), as they prefer drier upland terrain. Once common throughout central India, Kanha is the only place it survives today. When Project Tiger started here there were fewer than a hundred barasingha. But with good protection and grassland restoration, the number is now over 1,000.

◆ ◆ ◆

The cloud cover remained and the rains returned, but only periodically, allowing for many more park excursions over the next several days.

My guides and I covered almost all the park trails. I became familiar with certain places and soon had my favorite maidans to look for animals. I saw countless spotted deer, gaur, nilgai, boar, monkeys, and scores of birds. But the tiger and barasingha remained elusive. We saw their tracks and ran into other people who had seen them. However, I had no luck, until one afternoon, when lightning struck twice.

One of the places we had visited several times to look for animals was the Shravan Tal (lake), set in a wide maidan. It is a good place to observe wildlife and is also a part of Kanha legend. According to local lore, there was once a king who accidentally shot a young man named Shravan with an arrow one night, mistaking him for an animal. Shravan had gone to fetch water for his blind parents who were on religious pilgrimage. He was their only son. And when they heard the news, both died on the spot. The place has been known as Shravan Tal ever since.

My guide and I arrived there early one morning. A slight mist hung over the water. It was a uniform blanket, undisturbed by the still air, broken up only by the occasional clump of grass growing thick and tall close to the shore—a perfect hideout for a deer or tiger.

We looked down at the muddy shore for barasingha tracks. There were dozens of footprints in many shapes and sizes—large gaur and pig tracks mixed in with dainty spotted deer prints. My guide pointed out some larger ones. "These are barasingha," he told me.

"They are fresh tracks. The deer must be close."

Soon after he spoke, a herd of 10 exploded out of the grass cover and into the open and then paused to watch us. "They are really not afraid of people. They know they are protected here and people come to watch them all the time. Their main threat is the tiger," he added. "Also the jackals kill the fawns when they lay up in the grass."

We sat and watched for half an hour until the herd moved towards the woods. Then in another bolt of speed, they took off into the cover.

We reentered the woods ourselves, where we soon encountered one of the most lethal killers of the Indian forest—the *mahul* (*Bauhinia vahlii*)—a massive vine that grows up the side of trees, envelops them, and eventually strangles them to death. The dead tree eventually rots, leaving behind only the living casket—a mass of thick interwoven vines.

The tree in front of us was still alive, but the vines were already two feet thick and had fully enveloped the main trunk. Aggressive tendrils were climbing high into the canopy to complete the grisly task of enclosure. Eventually the vines will grow thick and fuse together to form an erect structure that endures after the host has died.

"Mahul is very common here," the guide explained. "The forestry people consider it a pest and want to kill it. But the local people like it. Especially tribals. They use the big leaves for plates and make medicines from other parts. Sometimes they sneak into the forest just to get them."

We left the mahul and continued our vain search for tigers. We drove around for hours, with no luck. We ran into another jeep that had spotted a leopard. But the tigers seemed to be elusive. The weather may have sent them far out into the forests.

Later that afternoon on our way out of the park, we came back to another maidan and were immediately greeted by a traffic jam. Five jeep loads of tourists were parked on the roadside, cameras clicking and fingers pointing. Children scrambled about to get a better view—and I knew what they had in their sights.

My driver sped up and came to a sudden halt right up next to them. There lay a tigress in the grass, about 20 yards off. She was unperturbed by the commotion and probably used to it. At Kanha, like Ranthambhore and Bandhavgarh, tour guides routinely use telephones or walkie-talkies to alert one another of a tiger sighting. Once a cat is spotted a small frenzy ensues as widely scattered vehicles converge (often at high speeds) to catch a glimpse of what they have traveled so far to see. That is why five jeeps were piled together here in this one place.

I was thrilled to see another tiger. But the atmosphere was unappealing—almost like being at a zoo. The muffled shrieks and low-level chatter finally became too much for the cat as well. She gave a low rumble of discontentment, got up, and returned to the peace and quiet of the forest. So did we.

◆ ◆ ◆

The rains picked up again the day before I left and continued through the afternoon into the night. Next morning it had stopped, but the air was heavy and the clouds hung low. It rained on and off during the entire drive back to

Jabalpur. And the rains intensified during the few days I stayed there before driving off to Kolkata, from where I then left India.

The monsoon had arrived. Soon the horizon would fill with more black and purple thunderheads, streaked by lightning and emitting low rumbles of thunder. The rains will then come in sheets and continue for days without pause, and the rains will continue throughout the entire summer. The parks will close, and the tiger will have the entire forest, now regenerated and teeming with life, to itself.

And perhaps someday the monsoon will regenerate even more forests, with more tigers, all across India and make them into the wild, magical places where tigers and people can live together as Rudyard Kipling once imagined they did.

— 10 —

A World of Tigers

Historically, tigers roamed all across Asia from eastern Siberia to the islands of Southeast Asia and west almost to the Mediterranean Sea. They occupied almost every habitat except deserts and frigid mountains. With the expansion of human population beginning about 10,000 years ago, tiger populations slowly became fragmented and isolated, leading to the emergence of several distinct tiger subspecies or races that varied in size, coloration, and habits.

The Indian tiger has always been the most common of all the various tiger races or subspecies, hence it is known as *Panthera tigris tigris*—the namesake of the entire species. But there are eight other subspecies that were all fairly common until the twentieth century. Since then these populations and their habitats, like that of the Indian tiger, have declined dramatically. Some races are already extinct, and others, on their way. No one really knows how many tigers survive in the world; estimates vary from less than 3,000 to 4,500.

Current population trends are not good. Habitat continues to vanish, and Asia's populations and economies continue to grow, placing ever greater demands on tigers and their habitat. Some predicted the tiger would disappear by the year 2000. That did not happen; the adaptable tiger still survives. But how long? And what is the story of world's other tigers?

The tiger that ranged furthest west—into present day Iran, Georgia, and Turkey—was the Caspian tiger (*P. t. virgata*). It was a large cat, only slightly smaller than the Indian tiger. Its short, thick pelt was paler than that of its Asian cousin, having brown rather than black stripes and a yellowish, not the deep orange, background.

The ancient Greeks knew of this curious striped cat from the east, and the Romans periodically caught them for display or for their infamous public animal fights, which sometimes included people. When Tamerlane, leader of the Mongol Empire, pushed west from the Asian steppe in the 1300s, he hunted Caspian tigers in a place called Lenkoran, in what is today northern Iran.

Caspian tigers had a precarious existence from the beginning. As a population, they were cut off from the rest of Asia's tigers by the vast plateaus and mountain ranges of central Asia. Originally this landscape was warmer and wetter, allowing the tiger to migrate there from Asia during the late Pleistocene—some 10,000 years ago. But as the climate grew colder and the landscape harsher, the animal became isolated.

The hardy tiger managed to survive in distinct habitats throughout this range, namely forested river valleys and the vast reed beds that lined central Asia's inland seas—Lake Balkhash and the Aral and Caspian seas. But people also desired these fertile areas and slowly began to transform them into farms.

This process accelerated greatly in the twentieth century, aided by modern technology and ideology. The eradication of malaria allowed the most impenetrable wetlands to finally be cleared and settled. Better machines and tools made this a rapid process.

But the tiger's fortunes really changed prior to World War II when the Soviet regime designated the Caspian Sea—the last large Caspian tiger stronghold—for development. The army was first sent in to kill off tigers to make human settlement safe. They were followed by engineers and farmers who cleared the reed beds and built roads and waterworks. In doing so they not only destroyed one of the world's most unique ecosystems but also ensured the tiger's demise.

Two cats were seen near Tamerlane's old hunting ground in the 1960s. They were the last. The Caspian tiger—the first tiger known to the West—is now presumed extinct.

Destruction and extinction also followed the tiger in the southeast corner of its range, in the islands of Java and Bali.

These lush tropical islands produced the smallest tiger species. The Bali tiger (*P. t. balica*) was less than eight feet long, including its tail, and it barely weighed 200 pounds. But in these tropical forests, where prey species are fewer and smaller, it was a suitable adaptation. Their population densities were naturally low because of their species-poor rainforest habitat. This did not help their survival.

The Bali tiger is a mystery. It was not known to Western science until the early twentieth century, and it was last seen in the 1940s in the forested mountains on the western side of Bali. There were never any zoo specimens and few photographs. Only a few bones and skins of it remain.

The demise of the Bali tiger was due mainly to habitat loss. Eastern Bali, with its rich volcanic soils, had been cleared for rice cultivation for centuries. The Dutch, who became the island's colonial rulers, later planted rubber plantations and cut timber in the remaining forested areas.

To their credit, the Dutch tried to protect the cat when it became endangered by creating the Bali Barat game reserve in the 1920s. But it was too late.

The small and widely dispersed population could not reproduce sufficiently, and the mysterious Bali tiger—the last known to science—was the first to become extinct.

Some biologists do not believe the Bali tiger was a separate race; rather, it was a Javan tiger. Since Bali is only a mile from Java, the cat probably crossed the channel and colonized the island. But, even if the Bali tiger is really a Javan tiger, it does not matter much as the Javan tiger is also extinct.

Unlike the mysterious Bali tiger, the Javan tiger (*P. t. sondaica*) was a common animal well into the twentieth century. And like its Indian counterpart, it was an intimate part of Hindu, animist, and Muslim culture—a forest animal that was worshipped and venerated.

Most Javan forest people believed in tiger spirits. Some thought their ancestors' souls lived on inside tigers; thus they would not harm the cat. The tiger, in turn, protected the people from harm. Others believed in spirit tigers or "were-tigers." The most important tiger spirits were present in the bodies of rulers, giving them courage and maintaining links between past and present.

Oddly, the Javan tiger never figured into the culture of the ruling Dutch the way the Indian tiger did with the British, especially with respect to hunting. Maybe the cats were too difficult to track in the dense jungles, or maybe Javan tigers were undesirable trophies, rarely reaching 300 pounds. But they were striking creatures with a deeper orange background color and narrower stripes than the Indian tiger. Most likely the tiger was not hunted much because the Dutch simply did not have a hunting heritage like the British—something that certainly benefited the tiger.

But the Dutch, like the British, did exploit Java's forests, and the Javan tiger eventually succumbed to human need. The island's growing population (one of the densest on earth) slowly pushed the cat into a few remote forests. The Dutch tried to protect the cat there by creating wildlife sanctuaries. But after independence, the drive for development led to rapid and more widespread forest destruction. Conservation had a low priority. Tigers were increasingly shot and poisoned, as they began to prey on village livestock to survive. Political unrest turned forest reserves into militant hideouts. The tiger population plummeted.

Twelve tigers were still confirmed to live in Java's largest sanctuary—Ujong Kulon—in the 1960s. But a decade later, their population had shrunk. Tracks of *P. t. sondaica* were last seen in the 1970s. The World Wide Fund for Nature launched a two-year study to confirm the tiger's existence in 1994. No tigers ever turned up. Perhaps only the spirit tigers remain somewhere in the deepest, darkest forest, unable to re-link past and present.

The most imperiled of the world's six remaining tiger races is the South China tiger (*P. t. amoyensis*). This is particularly ironic, as the tiger is believed to have originated in China during the Pleistocene and migrated from there to all other parts of Asia.[1]

The South China tiger's natural range was the temperate and semitropical forests of southern and eastern China. The cat was still common in the more remote areas of its former range until the middle of the twentieth century. Some 3,000 to 4,000 still survived until the late 1950s. But Mao Zedong's ambitious campaign to modernize the countryside (as discussed in chapter three) eradicated this striped "pest" as effectively as the Soviets' efforts did along the Caspian Sea. After Mao's experiment ended, the demand and desire for resources and farmland did not, thus worsening the Chinese tiger's plight.

In the 1990s the Chinese government claimed that some 100 tigers remained in the southern part of the country in the mountains of Jianxi Province. But most Western biologists think this is highly inflated. The real number is fewer than 50, perhaps as low as 20. What is worse is that they live in three isolated populations, making breeding difficult. The chances for survival are slim, and it may be the first tiger species to become extinct in the twenty-first century.

The Chinese government has made trading tiger bone and pelts a crime—sometimes even punishable by death. But the illegal trade and demand for tiger parts continues. This has prompted the Chinese government to come up with its own solution—tiger farming. Over 4,000 tigers—more than the entire wild population—are raised on special farms across China. They are bred for zoos and public display, but especially for their parts—which the government claims are only harvested after the animals die a natural death (although critics dispute this).

China has been pushing the idea of tiger farms as a way to reduce illegal poaching of wild tigers. They claim that they can produce 100,000 tigers per year, which would supply the entire market for tiger parts. It would also depress prices, thus reducing the incentive to poach.

This idea has created enormous debate in the conservation community. Some believe the commercialization of tiger parts is the best way to reduce poaching. Others claim this will have the exact opposite effect. More tiger parts on the market will create more interest and demand for them, leading to even greater pressures on wild tigers. The issue has yet to be resolved.

The Sumatran tiger (*P. t. sumatrae*) is next on the endangered list. It is the only remaining jungle island cat, similar in size and color to the Javan. Fortunately it lives on the much larger and less populated island of Sumatra, which has helped it survive. But it shares the same demographic problem of small populations scattered over vast areas of rainforest.

This was not an issue until the human population of the island began to grow after World War II. Historically, people were few on Sumatra. The island was remote, had mountainous terrain and poor soils. But Indonesia's growing economy and population needed land and resources. The island was found suitable for rubber and oil palm plantations and had abundant timber. It also had oil. As a result Sumatra's human population swelled, and its forests shrunk.

Poachers also began to exploit this large and virtually untapped tiger population beginning in the 1960s. Hundreds of pounds of bone left Sumatra every year for over three decades. Most went to the newly industrializing countries (NICs) of Asia—South Korea, Taiwan, and Hong Kong—which, ironically, are also known as the "Asian Tigers" because of their aggressive economic growth.

By the end of the 1970s it was thought that about 1,000 Sumatran tigers survived. Today the number is believed to be half that, perhaps as low as 200. There are about a dozen forest tracts with tigers remaining on Sumatra. Most are under protection, but only five are of significant size, and none are contiguous.

The hope is that two or three of the largest tracts can be adequately protected to support significant breeding populations. But, like in India, all these reserves are surrounded by villages and farms, making conservation a particularly daunting task and one requiring local cooperation to work.

Another threatened tiger is the Amur tiger (*P. t. altaica*), once known as the Siberian tiger. There are 400–500 Amurs left in the unpopulated larch and coniferous forests of eastern Russia and the adjoining border regions of China and North Korea.

These few hundred cats are spread out over tens of thousands of square miles because of low prey densities in these frigid forests (even lower than those in rainforests). Despite a low population, the story of the Amur tiger is one of the few tiger conservation successes.

Protection of large areas of land, combined with strong anti-poaching efforts and consistent scientific analysis and monitoring has helped this species to almost double its population in the last two decades.[2] There are also over 1,000 Amur tiger in zoos all over the world, making it one of the most common zoo tigers in the West.

The Amur is the world's largest cat, reaching 14 feet including the tail and almost 700 pounds (although Indian tigers from the Himalayan foothills are sometimes heavier, but with a smaller frame). The Amur is known for its heavy pelt and pronounced white underside and throat. Older males grow thick fur around their necks resembling a mane.

The deep orange and pronounced black stripes would seem to be out of place in the tiger's snowy domain. But the combination of striped body and stark white underside allows the cat to blend perfectly into a snowy landscape studded with stands of larch and conifers. It is another example of the amazing adaptability of the tiger.

Because it lives in some of the world's most inhospitable habitats, the Amur has been relatively safe from poaching. The greater threat has been habitat loss driven in recent years by the demand for oil and timber by a cash-hungry Russian economy.

The Amur tiger was still common throughout eastern Russia, Manchuria, and Korea until the early twentieth century. The last one seen in South Korea was in the 1940s, and populations in North Korea and China (Manchuria) dwindled thereafter. The two main Amur tiger populations remaining are in the Primorsky and Khabarovsk forests, both of which, fortunately, have large areas of contiguous forest, which this cat—more than any tiger species—needs to survive.

The Indochinese tiger population is the largest after the Indian—500–1,000 cats. It ranges throughout Southeast Asia from Myanmar to Vietnam and south to the tip of Malaysia. But its population is probably the hardest to verify of any tiger species, because it covers a vast area spread over six different countries.

For years it was considered either an Indian tiger, in the western part of its range, or a South China tiger, farther north. Even today biologists debate whether the Indochinese is really a separate race, as it is hard to distinguish from its Indian or Chinese neighbors. But there are discreet differences—a deeper orange background and more numerous, narrower stripes. The debate continues.

The Indochinese was not recognized as a separate race until 1968, and it was named in honor of the famous Jim Corbett—*Panthera tigris corbetti*—a decade after Hailey National Park in India was changed to Corbett National Park.

The largest Indochinese population is believed to be in Myanmar, which still has sizeable swaths of undeveloped forests with ample prey species. But fewer than 5 percent of these forests are protected. There are also political problems. The country's rigid dictatorship is riddled with corruption. As a result, poaching and smuggling of tiger parts and other wildlife contraband is rampant—much of it organized by government officials. The fact that some of Myanmar's best tiger forests border China is not reassuring.

Another population center for *P. t. corbetti* is Indochina. Decades of war and strife have isolated this part of the world from scientific study. The impact of this conflict on tigers is not known. War prevented many forests in Vietnam, Laos, and Cambodia from being developed for timber and agriculture. But war also forced people into the forests to hunt animals simply to survive.

Even today local game hunting is a major conservation issue in the region. Poor people trap, snare, spear, and shoot almost any animal—from insects to elephants—simply to get protein. Even when tigers are not directly killed, they suffer because their prey base has vanished. Many of Indochina's forests are still beautiful and intact. But many are also lifeless because of rampant local hunting.

The largest tiger forests remaining in Indochina are in the south—the forested hill tracts of northern Cambodia and the tri-state border region just to the east. Another parcel is in the north along the Vietnam-Laos border. These are contiguous forests that could support hundreds of tigers, but a reduced prey base has probably greatly diminished that potential.

In Thailand the tiger has received protection after years of poaching, much of it to fill the demand of wealthy Thais for tiger rugs and stuffed tigers—a sign of newfound prosperity. In the 1990s the government became more vigilant in response to the rapid loss of tigers. Today several reserves exist throughout the country. Currently western Thailand has the best remaining forests, with several sanctuaries that protect contiguous forest habitat. The overall tiger population in Thailand is a few hundred cats.

Malaysia also has a few hundred cats, found in the center of the mountainous Malay Peninsula. The greatest threat to these tigers is not so much poaching but the government and its aggressive policy of expanding timber harvesting and the plantation economy to fuel economic growth. The country has the dubious distinction of having maintained one of the highest deforestation rates in the world over the last two decades.

In 2004 these Malaysian tigers, especially those from the southernmost tip of the country, were named a new subspecies based on genetic differences from the Indochinese race. These cats are also smaller, closer in size and color to the Sumatran tiger.

The Malaysian tiger was named *Panthera tigris jacksoni*, in honor of the British journalist Peter Jackson who spent his career writing about tigers and organizing efforts to protect them. The Malaysian tiger population is estimated to be anywhere from 300 to 800 cats.

The Indochinese and Malaysian tigers may benefit from the growing prosperity in Malaysia, Vietnam, and Thailand in particular. More money and an educated populace are slowly making conservation a political issue. Like with India, these countries may become safer tiger havens in the future. Southeast Asia also has the potential to make conservation an area of international cooperation, because so many of its tiger forests exist on shared borders. For tiger protection to be effective it must become a common regional political goal.

The Indian tiger (*P. t. tigris*) still represents the world population. The majority—about 1,500—are found in India proper. A few hundred more reside in neighboring Bangladesh (in the Sunderbans), Nepal, Bhutan, and western Myanmar.

Two recent developments have benefited tigers in South Asia. One is the reduction of violence in Nepal, which has been in civil war for years. During the struggle, which was largely rural, most wildlife sanctuaries could not be monitored. So tigers were poached and habitat was destroyed.

In neighboring Bhutan, the government has created a new social philosophy, one that seeks to counter the dominant materialism of much of the world. Rather than gross national product as the measure of national success, the government has pushed a new idea of "gross national happiness," with a focus on spiritual and nonmaterial markers of national success. Such an approach would naturally have beneficial effects on nature and people's relationship to nature, all of which would only help Bhutan's small tiger population of a few dozen animals.

India remains the best hope for tiger conservation. Despite all its problems, it is still a model for tiger conservation. Not only does it have more distinct tiger habitats with larger tiger populations than any place on earth, it also has the world's second largest contiguous bloc of forest—the terai arc (which also includes Nepal and Bhutan). Only the Russian Far East is larger. Biologists see this particular tiger landscape, with its high human populations mixed with abundant forests and wildlife, as the basis for successful tiger conservation in other countries.

◆ ◆ ◆

In 2005, the world's preeminent tiger conservation organizations produced the most thorough tiger conservation proposal to date, entitled *Setting Priorities for the Conservation and Recovery of Wild Tigers: 2005–2015*,[3] or simply *Wild Tigers*. This document, designed to be constantly revised as situations change, is also global in scope. It is the first to make a comprehensive assessment of the tiger's plight and explore comprehensive means to resolve it.

The plan emphasizes more research and surveillance, integrating tiger conservation into regional and national development plans, the curtailing of poaching, and the recruiting of tiger spokesmen and women of stature who can make tiger conservation a topical issue.

The main focus is protecting and restoring remaining tiger habitat, which has declined by almost 50 percent worldwide in the last 15 years. The total amount of tiger habitat left in the world is now only 5 percent of the animal's original range.

This bold, new strategy builds upon the older protected area system that India helped pioneer through Project Tiger. This system created a network of protected areas that were representative of the animal's various natural habitats (biomes). *Wild Tigers* envisions establishing such a network across all of Asia. The key difference is that it expands the older core-periphery idea into an even larger conservation designation—the concept of tiger ranges or Tiger Conservation Landscapes (TCLs). These are habitats that incorporate wilderness areas, human use zones, and, most importantly, corridors that link tiger habitat into contiguous swaths to facilitate breeding, hunting, and movement, thereby ensuring the long-term survival of large cat populations.

TCLs are defined as areas with at least five tigers that have lived and bred there for 10 years. The study identified 76 TCLs covering almost 500,000 square miles. These are further divided into distinct classes based on the quality of tiger habitat. Class 1 TCLs have over 100 cats; class 2 at least 50; class 3 at least 5 cats; and class 4 are potential tiger habitats, but with no population data. The study also identified an additional 300,000 square miles of viable tiger habitat (representing over 500 different forest landscapes) that may still have tigers or where tigers are absent but could be reintroduced.

Of all 76 TCLs, only 16 are class 1. Fifteen are class 2, with the remainder divided between the last two classes. But the good news is that half of all TCLs could support over 100 tigers, and at least seven class 1 TCLs could support over 1,000 tigers. Together with the smaller remaining TCLs and other viable forest habitats, the world tiger population (with good conservation) could reach 15,000 animals or higher.

Where are the TCLs found? Naturally, the majority—40—are in India and neighboring Nepal, Bhutan, Bangladesh, and western Myanmar. Next is Indochina with 20, followed by Southeast Asia and Indonesia with 15, and there are 2 in the Russian Far East.

The study identified 20 TCLs as "global priority" habitats. These are the largest and most promising tiger habitats, where tigers have the best chance of long-term survival. Of these, 11 are in India and adjacent countries; the other 9 are spread throughout Indochina and Southeast Asia.

The long-term tiger conservation goal is to establish as many class 1 Tiger Conservation Landscapes as possible by protecting them and elevating others to that level. The only way for this to happen is if tiger conservation becomes a coordinated effort involving international organizations, national governments, and local communities.

This approach represents a real change in conservation thinking and practice. It goes beyond the older "defensive" strategy, which relied on strictly defined and protected tiger sanctuaries, to one that seeks to integrate itself into the political, economic, and cultural life of a nation and people.

The struggle now is to restore the tiger to its proper place in the ecosystem as the supreme forest predator—the King of the Jungle—and as a symbol of national and cultural pride. And only people can make that happen.

Notes

CHAPTER 1

1. The tiger migrated west and north from what is today China during the late Pleistocene, as recently as 10,000 years ago. There is still debate over where in China (including Mongolia) the cat first originated.

CHAPTER 2

1. See François Bernier, *Travels in the Mogul Empire, A.D. 1656–1668* (London: A. Constable, 1891).
2. All figures are from Kailash Sankhala, *Tiger! The Story of the Indian Tiger* (New York: Simon and Schuster, 1977), 132–133.
3. K. Anderson, *Man-Eaters and Jungle Killers* (London: Allen and Unwin, 1957), 7–8.
4. K. Anderson, *Nine Man-Eaters and One Rogue* (London: Allen and Unwin, 1957), 108–109. Anderson was known for his dramatic descriptions, which some critics believe were slightly embellished. But his record of shooting man-eaters and other jungle killers is not disputed.

CHAPTER 3

1. *One Man and a Thousand Tigers* (New York: Dodd, Mead and Co., 1959) was the title of a famous hunting book by Colonel Kesri Singh. He, together with his clients, shot at least that many cats in the Rajasthan environs over a lifetime.
2. More modest estimates place the number at 200,000–300,000, most of which are found in Africa.
3. Gujar or Gujjar peoples are found over much of India. They are frequently nomadic and follow Hindu, Muslim, or Sikh religions.

CHAPTER 4

1. The original nine reserves included Corbett, Manas, Sunderbans, Simlipal, Palamau, Kanha, Ranthambhore, Melghat, and Bandipur.
2. Some biologists believe that this number was already inflated; 3,000 may have been a more accurate figure.
3. Gandhi was assassinated by her Sikh bodyguards, who were active in the freedom movement for the state of Punjab—the traditional Sikh homeland.

CHAPTER 5

1. In contrast, the rich topsoils of the American Midwest or central California are only a few hundred feet deep.
2. Academics debate the veracity of the Aryan invasions. Some claim theirs was a slow migration, more peaceful than warlike. Others see them as an outside force that imposed their will on native peoples.
3. Muslims in other tiger range countries, especially Southeast Asia, also place religious significance on tigers. Some see killing a tiger as an offense to Allah, while others believe Allah allows tigers to kill people as a punishment for sins.
4. M. Twain, *Following the Equator: A Journey Around the World*, Author's National Edition, vol. 6 (New York: Harper and Bros., 1899), 174.

CHAPTER 6

1. This war led to the independence of East Pakistan, which became Bangladesh.
2. Other duars sanctuaries with tigers include Chapramari, Neora Valley, and Gorumara. Altogether, these sanctuaries may have two dozen cats at most.
3. The maharaja himself shot 365 tigers in a 30-year period beginning in 1871. That is almost 10 times the number of cats remaining in all of the duars.
4. The dry interior prevented the teak from establishing itself naturally in the northeast.
5. The Global Environmental Facility is an organization that pools funds from wealthier nations and disperses them to developing nations under the direction of the United Nations and the World Bank. It funded several eco-development programs throughout India from 1996 to 2001, including Ranthambhore and Kalakad-Mundanthurai.

CHAPTER 7

1. Some anthropologists believe that the Negritos were the world's original human inhabitants and were widely dispersed throughout Asia and Africa before being overwhelmed by other more populous peoples.
2. The name "Gondwanaland" was coined by Austrian geologist Eduard Suess in reference to the Upper Paleozoic and Mesozoic rock formations found throughout central India—the land of the Gonds.
3. Two of the most deadly Indian vipers are Russell's viper (*Daboia ruselii*) and the saw-scaled viper (*Echis carinatus*).

4. The Indian bullfrog (*Rana tigerina*), which measures 15 inches with legs extended, is the favorite. But smaller species are increasingly taken as the bullfrog population declines.

5. J. Connell, "Wild Dogs Attacking a Tiger," *Journal of the Bombay Natural History Society* 44 (1943): 468–470.

CHAPTER 8

1. François Bernier, *The Empire of the Great Mogul* (London: Moses Pitt, 1672).

2. R. G. Burton, "Man-Eating Tigers on Saugur Island in the 18th Century," *Journal of the Bombay Natural History Society* 27 (1920): 386. Burton quotes this story that was originally published in the *Annual Register* of Calcutta, October 12, 1787.

3. K. Chakrabarti, *Man-Eating Tigers* (Calcutta: Darbari Prokashan, 1992), 76.

4. The crocodiles are still locally common in Australia and New Guinea.

CHAPTER 9

1. J. Forsyth, *The Highlands of Central India: Notes on Their Forests and Wild Tribes, Natural History, and Sports* (London: Chapman and Hall, 1871).

2. Interestingly, Kipling's forest experiences were confined to the Himalayan foothills. He undoubtedly pieced together tales from the many people he knew who had traveled to central India.

3. Kanha is one of the cases where villages willingly resettled outside the park and have maintained an amiable relationship with the government ever since.

4. The deer include the spotted deer, barking deer, sambar, hog deer, and barasingha. The antelopes are the nilgai, blackbuck, and chinkara.

5. R. Kipling, *The Jungle Books, Volume One* (Garden City, NY: Doubleday and Co., 1948), 103.

CHAPTER 10

1. There are several theories for tiger evolution and dispersion. The main arguments are over what part of China (including Mongolia) the cat first evolved and the migration routes it took.

2. Most of these projects are joint Russian and American operations—a fitting conclusion to years of Cold War animosity.

3. E. Dinerstain, C. Louks, A. Heydlauff, E. Wikramanayake, G. Bryja, J. Forest, J. Ginsberg, S. Klenzendorf, P. Leimgruber, T. O'Brien, E. Sanderson, J. Seidenstecker, and M. Songer, *Setting Priorities for the Conservation and Recovery of Wild Tigers: 2005–2015* (Washington, D.C./New York: World Wildlife Fund, Wildlife Conservation Society, Smithsonian, and the National Fish and Wildlife Federation—Save the Tiger Fund, 2006). Other donors and supporters besides the four main publishers include the United National Foundation, the U.S. Fish and Wildlife Service, ZSL—Living Conservation, and Exxonmobil.

Index

About the Author

TOBIAS J. LANZ teaches international and environmental politics at the University of South Carolina. He has degrees in wildlife science (BS) and agricultural economics (MS) from Texas A&M University, an MA in international relations from San Francisco State University, and a doctorate in international studies from the University of South Carolina. He did his dissertation fieldwork on rainforest conservation at Korup National Park in Cameroon, Africa in 1994 and 1995.

Professor Lanz has been studying the social and natural history of Indian tigers since 1998. He has traveled widely across the subcontinent and has visited and conducted field research at most of India's premier tiger sanctuaries.

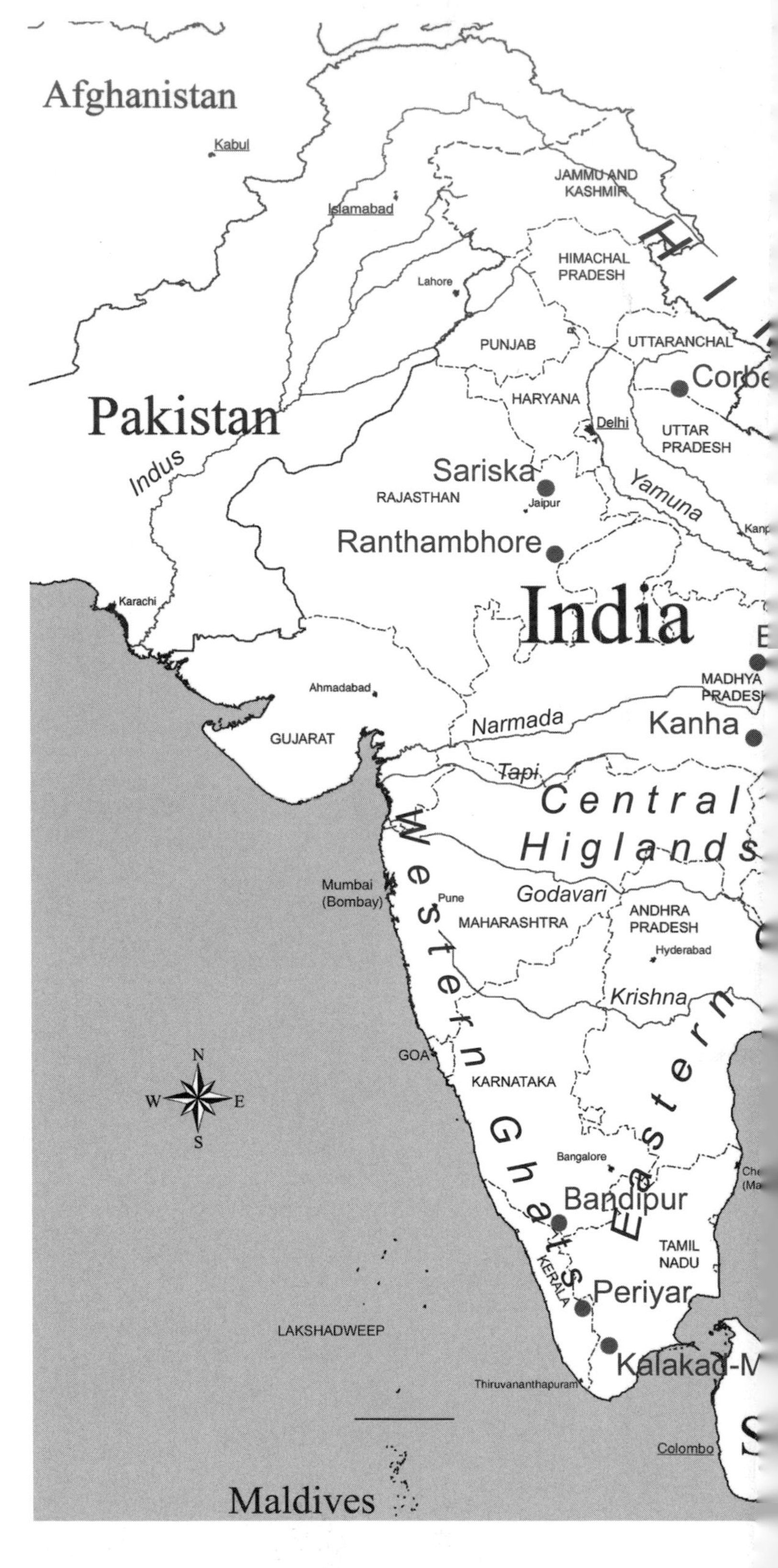
Afghanistan
Kabul
Islamabad
JAMMU AND KASHMIR
HIMACHAL PRADESH
Lahore
PUNJAB
UTTARANCHAL
HARYANA
Delhi
UTTAR PRADESH
Pakistan
Indus
Sariska
RAJASTHAN
Jaipur
Yamuna
Ranthambhore
Karachi
India
MADHYA PRADESH
Ahmadabad
Narmada
Kanha
GUJARAT
Tapi
Central
Higlands
Western Ghats
Mumbai (Bombay)
Pune
Godavari
MAHARASHTRA
ANDHRA PRADESH
Hyderabad
Krishna
Eastern
GOA
KARNATAKA
N
W
E
S
Bangalore
Bandipur
TAMIL NADU
KERALA
Periyar
LAKSHADWEEP
Thiruvananthapuram
Colombo
Maldives